THERMAL

JOSEPH GIACOMIN

THERMAL

SEEING THE WORLD
THROUGH 21ST CENTURY EYES

PAPADAKIS

To Andreas Papadakis, whose vision and spirit made this book possible

First published in Great Britain in 2010 by
Papadakis Publisher

PAPADAKIS

An imprint of New Architecture Group Limited

HEAD OFFICE: Kimber Studio, Winterbourne, Berkshire, RG20 8AN
DESIGN STUDIO & RETAIL: 11 Shepherd Market, Mayfair, London, W1J 7PG

Tel. +44 (0) 1635 24 88 33
Fax. +44 (0) 1635 24 85 95
info@papadakis.net
www.papadakis.net

Publishing Director: Alexandra Papadakis
Editor: Sarah Roberts
Design Director: Aldo Sampieri

ISBN 978 1 901092 84 4

Copyright © 2010 Joseph Giacomin and Papadakis Publisher
All rights reserved
Joseph Giacomin hereby asserts his moral rights to be identified as author of this work.

No part of this publication may be reproduced or transmitted in any form or by any means, electronic or mechanical, including photocopy, recording or any other information storage and retrieval system, without prior permission in writing from the Publisher.

Printed and bound in China

page 2: A woman performs the elegant movements of a traditional Malaysian dance

page 3: A large crowd of people cluster together as they enter Wembley Stadium for a music concert

opposite: A crowd gathers outside an underground station with the London Eye in the background

pages 6-7: Groups of people relaxing and enjoying the gardens in front of the Natural History Museum, London

pages 8-9: The engine, brakes and exhaust system of this bus all generate heat

pages 10-11: Students in a workshop

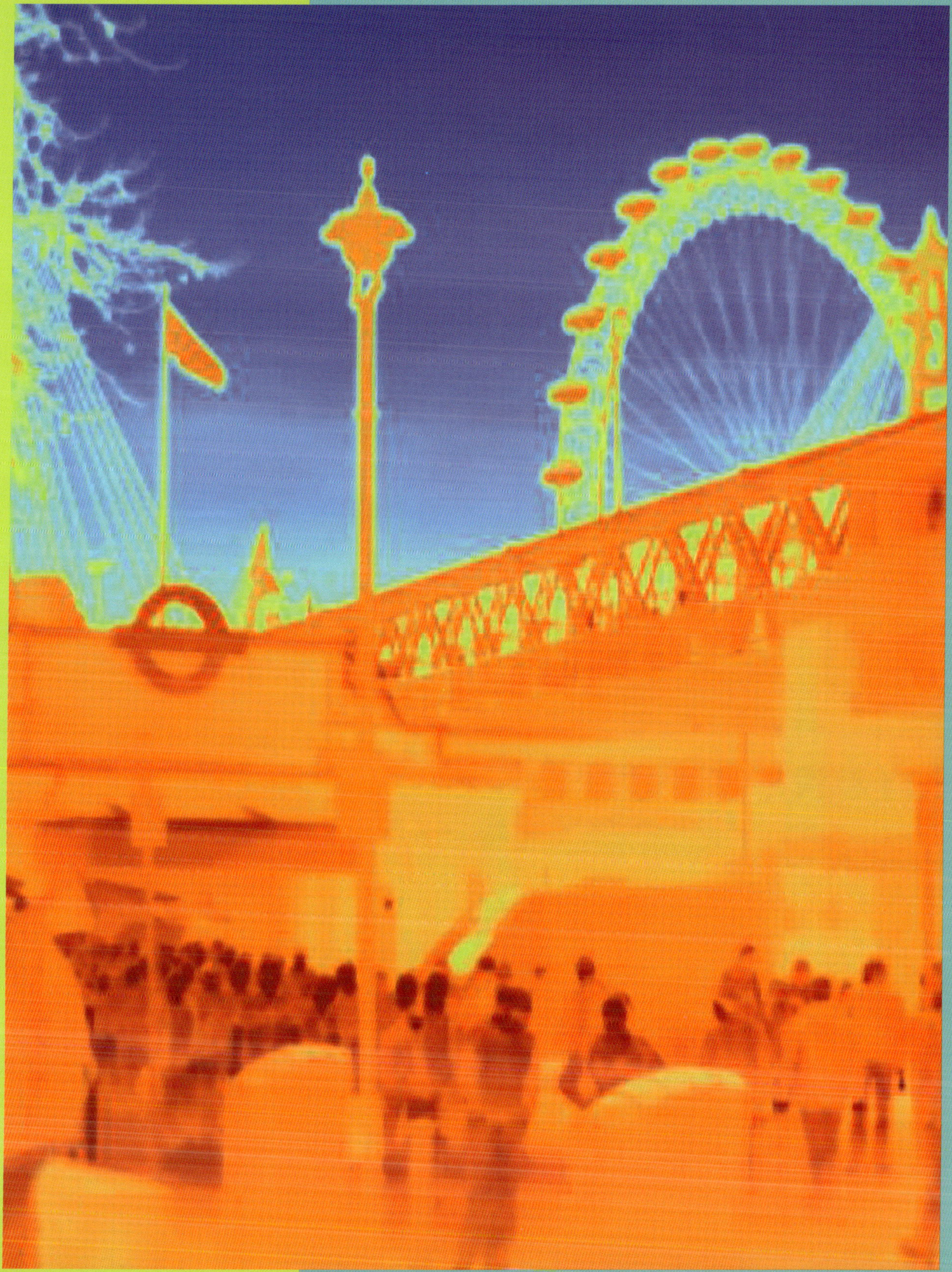

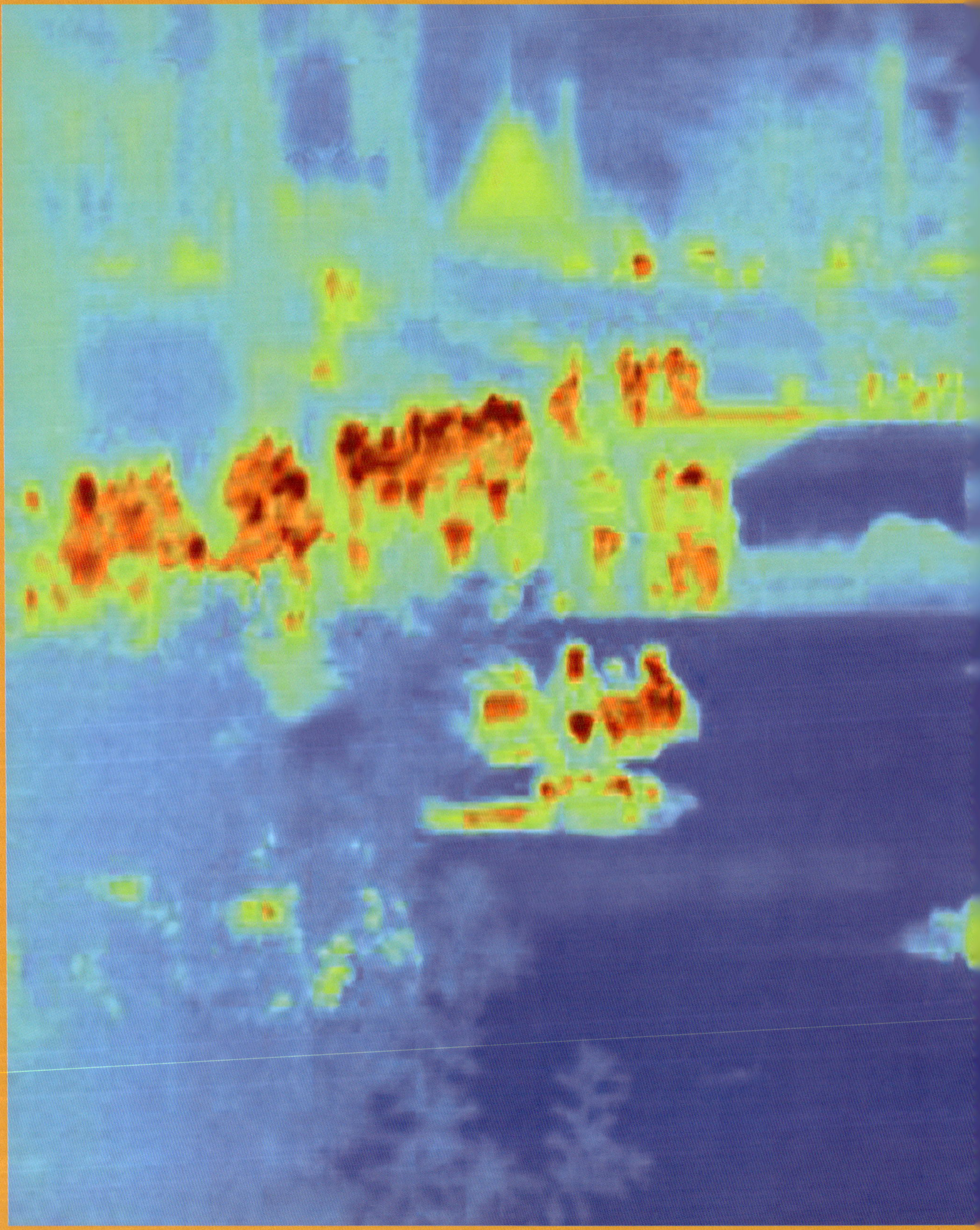

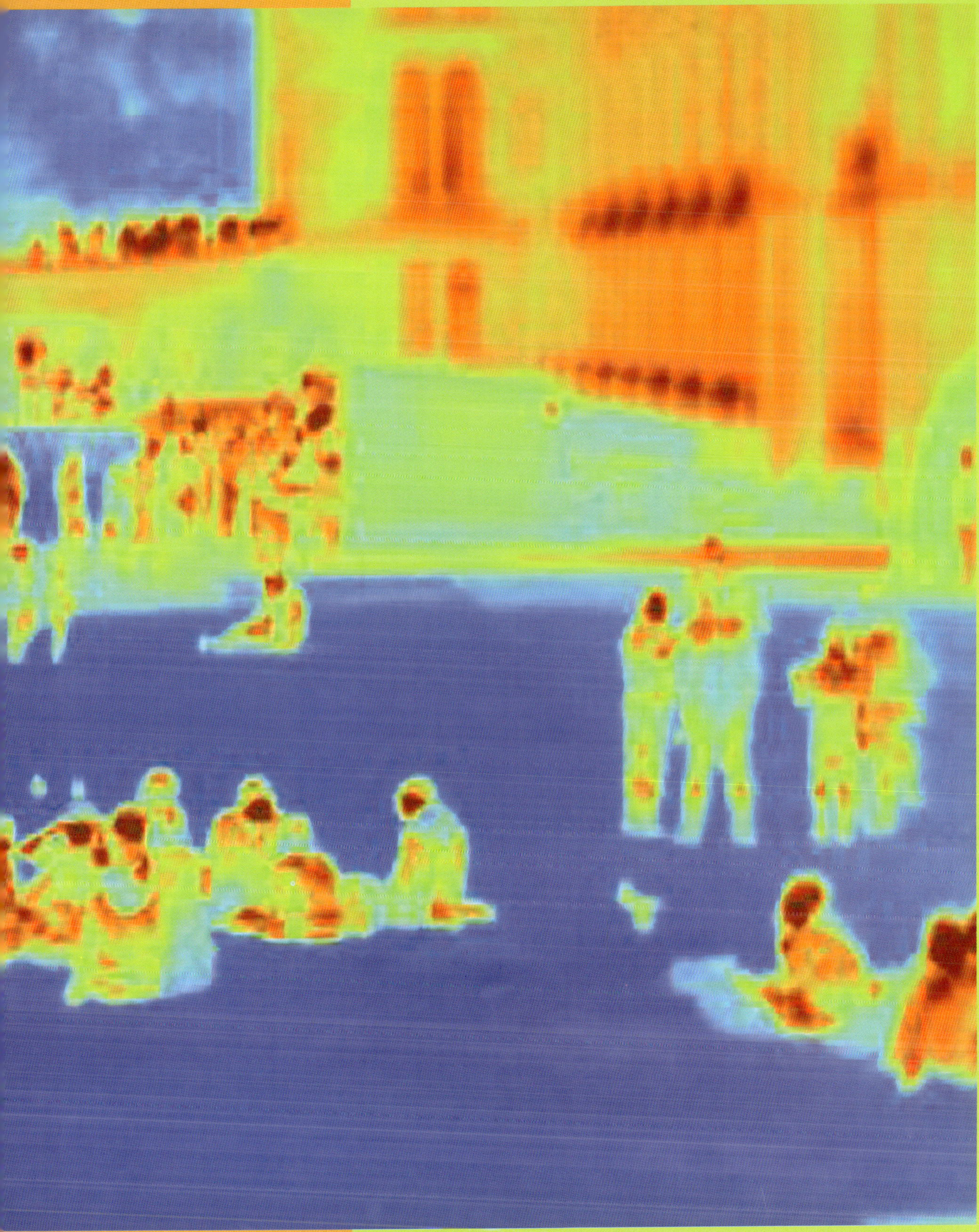

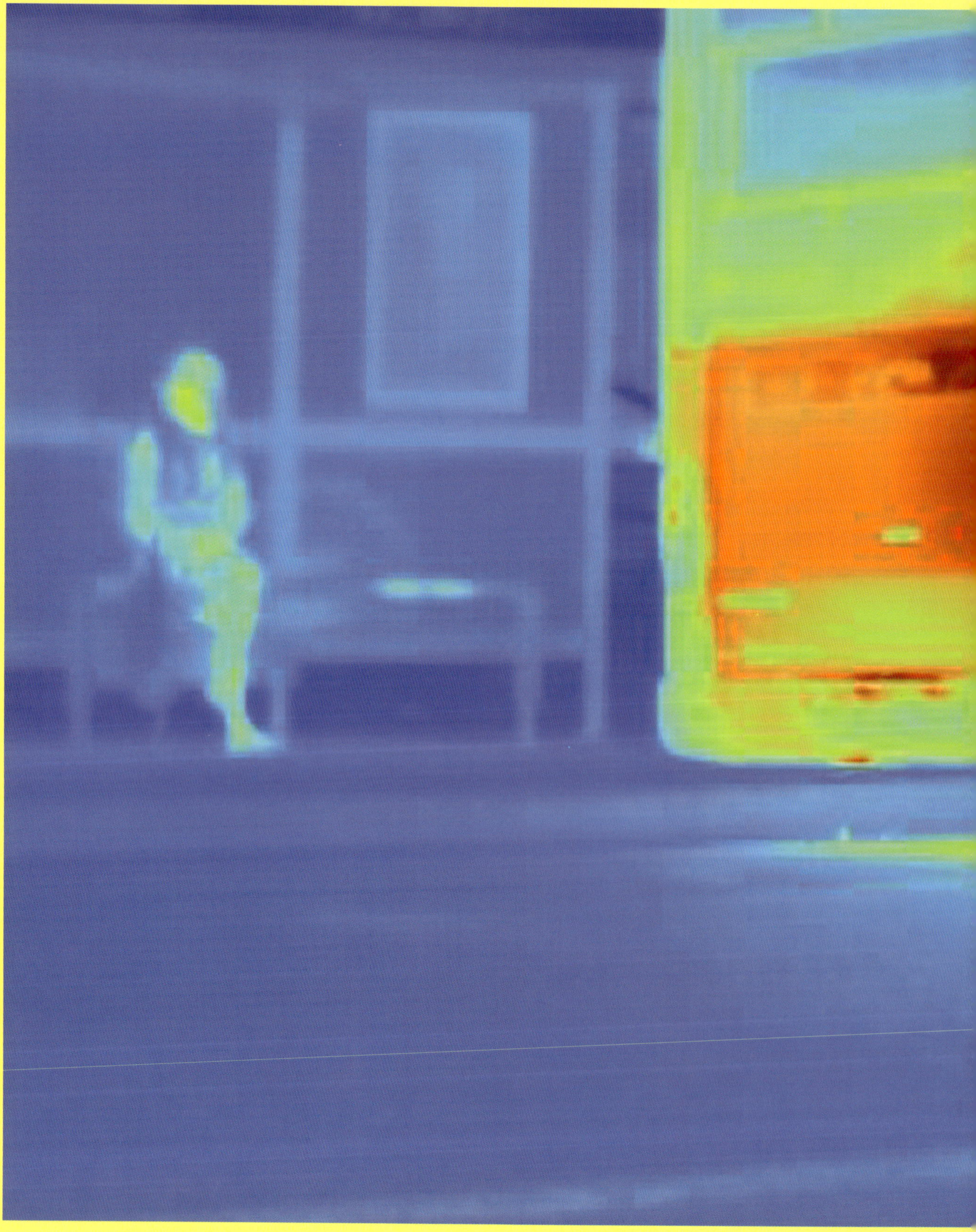

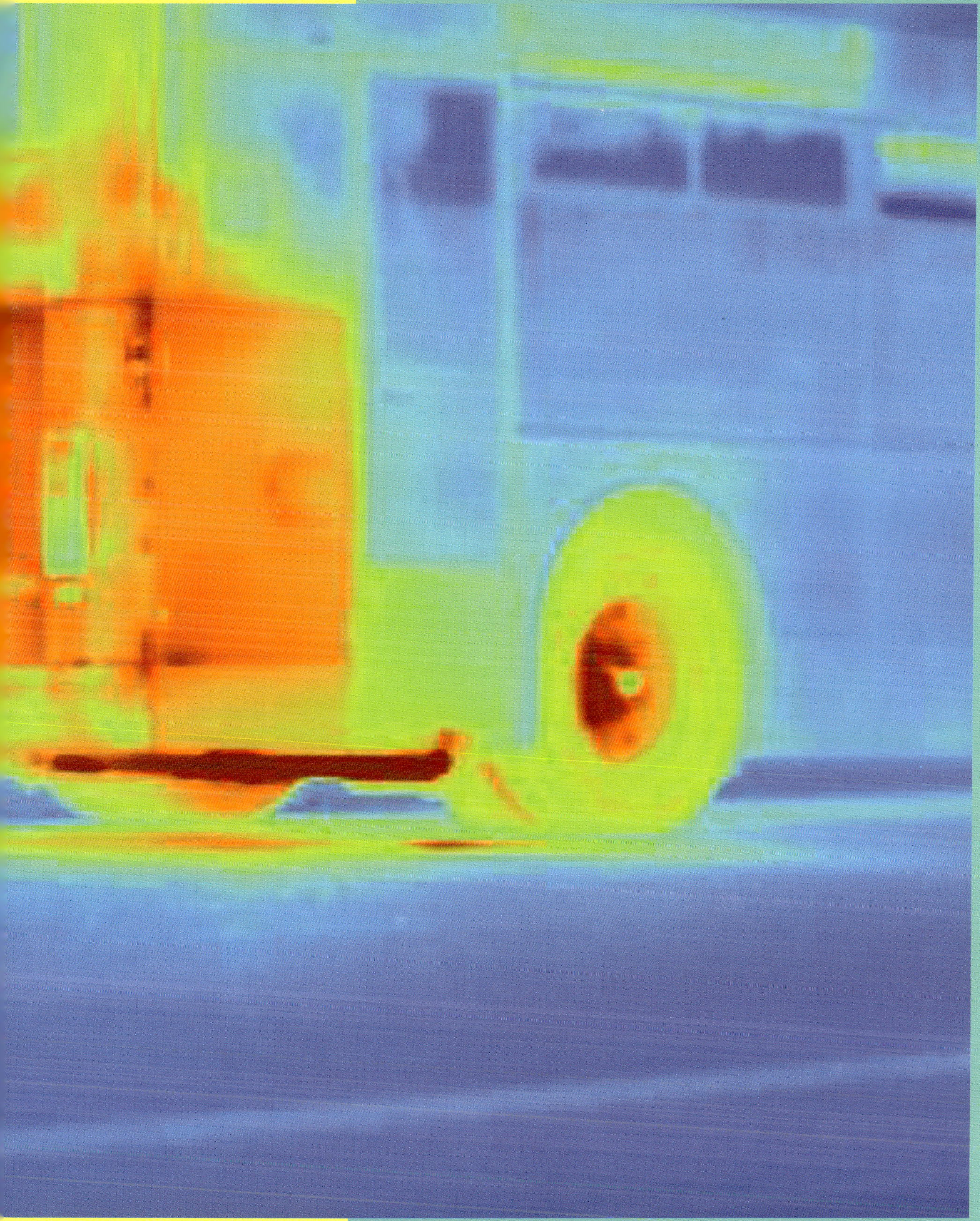

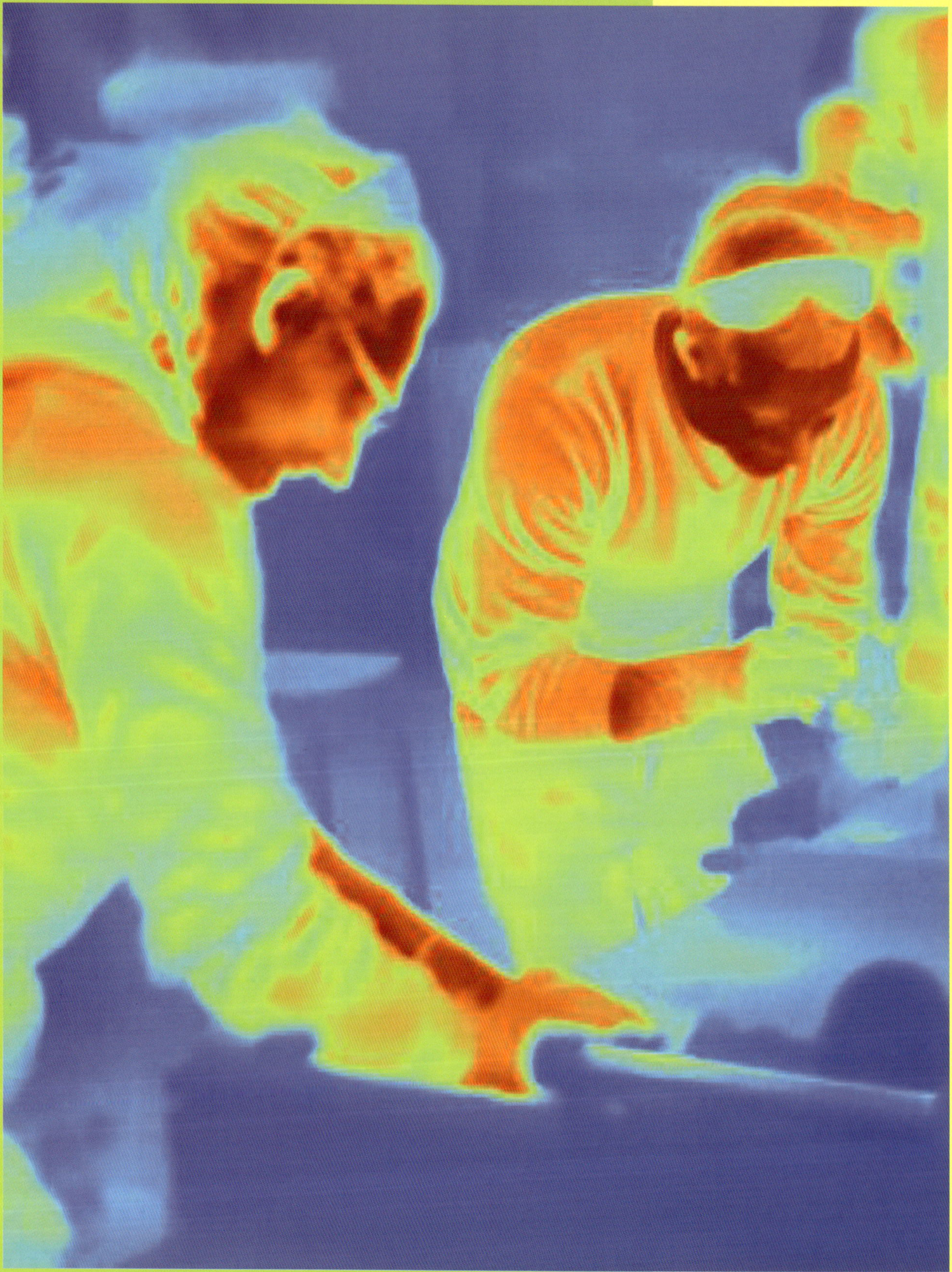

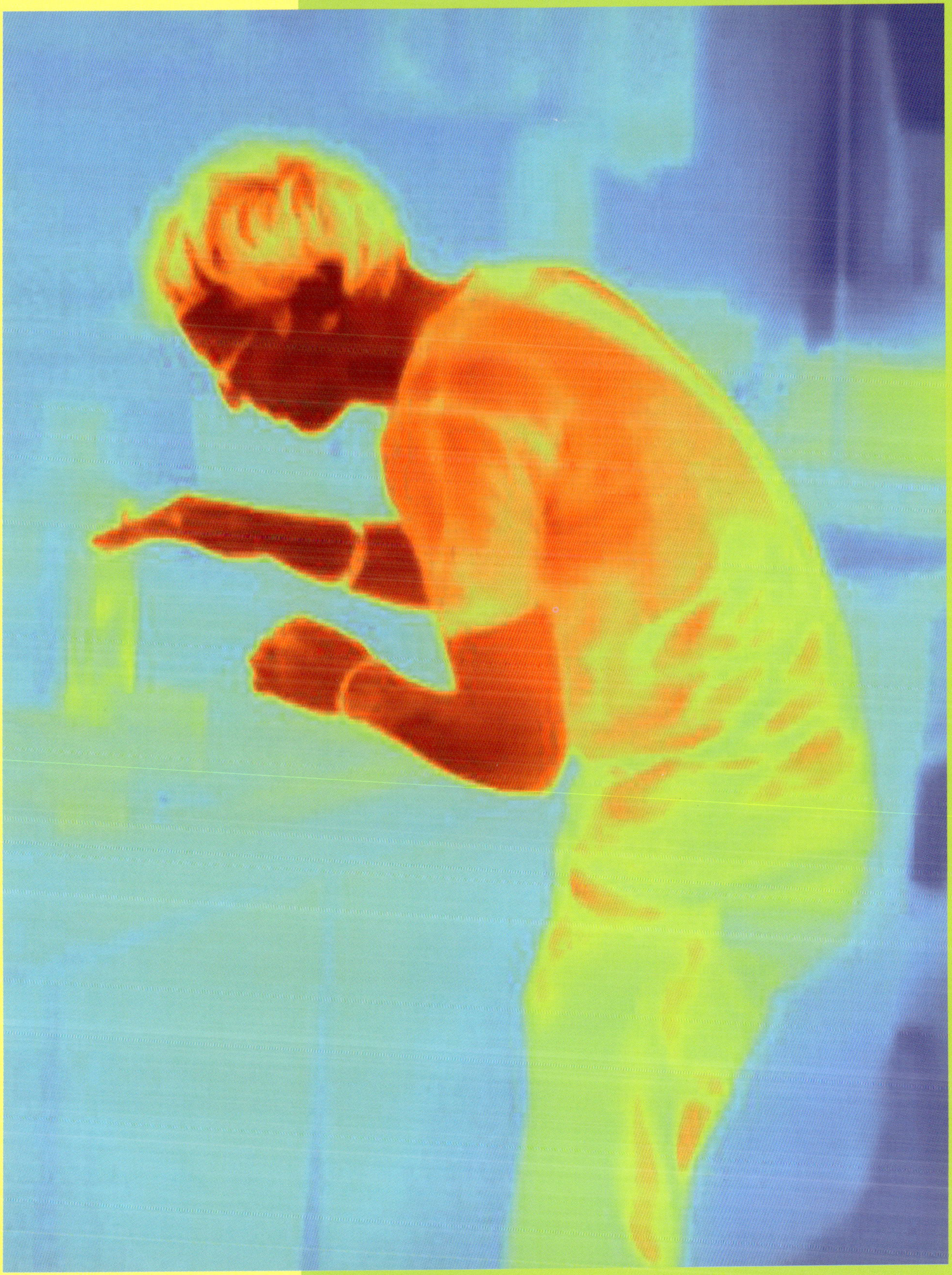

CONTENTS

15 INTRODUCTION

21 WHAT IS PERCEPTION?

69 PSYCHEDELICS

93 PERCEPTION ENHANCEMENT

105 GUIDE TO THERMAL IMAGES

112 BIBLIOGRAPHY

22	Global Warming	54	Salisbury Cathedral
26	Electricity	56	London
28	Heat and Fire	60	Millennium Bridge
30	People and Animals	70	Underground Station
32	Motion	72	Underground Trains
34	Sunlight	74	Vehicles
36	Clothing	76	Churchyard
38	Lighting	78	Stonehenge
40	Materials	80	Benches and Chairs
42	Water	82	People
46	Thermal Exchange	86	Hair Dryer
48	Visible vs Thermal	89	Glasses
50	Sky		

14

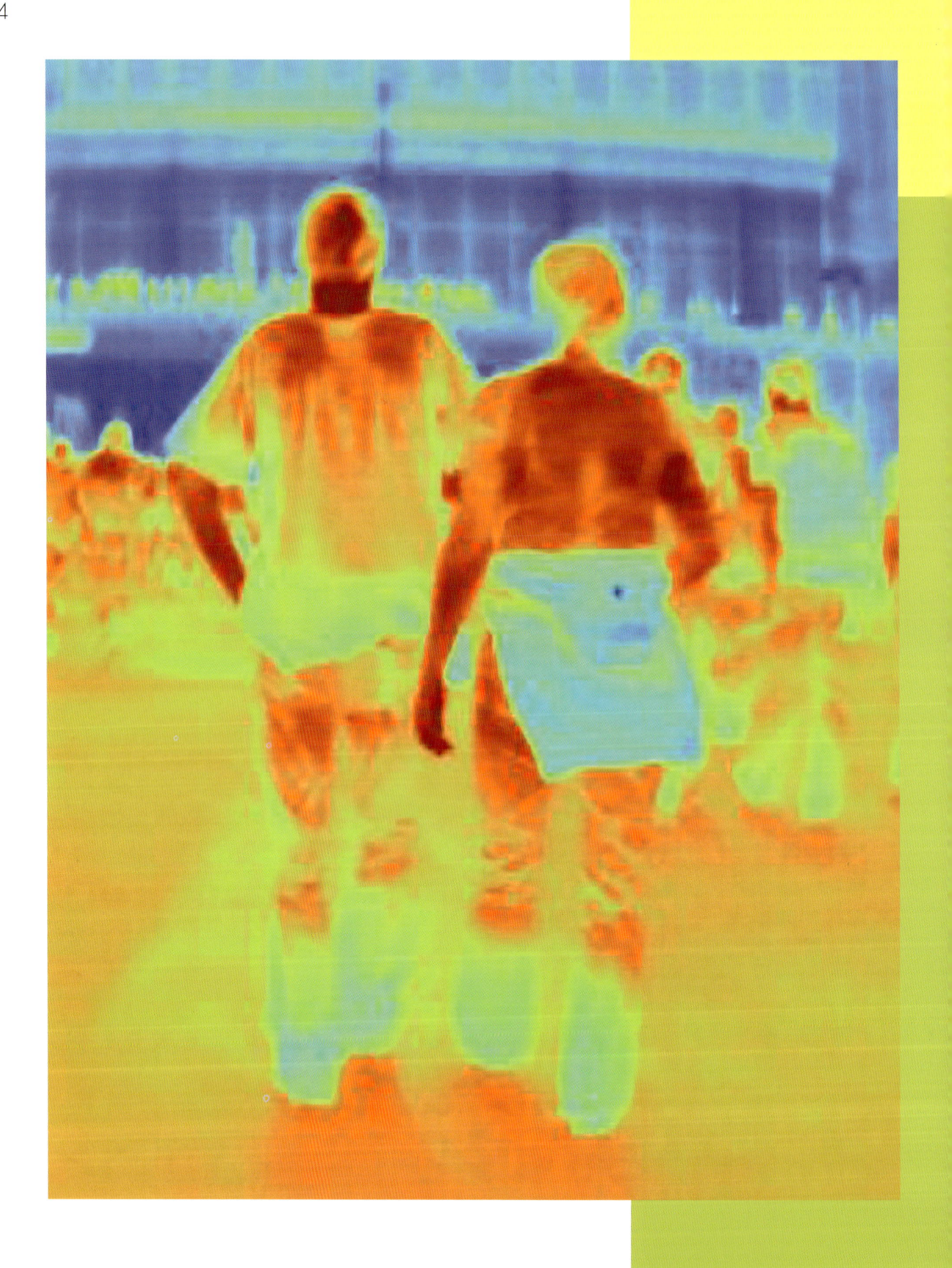

INTRODUCTION

The contents of this book revolve around the topic of perception, a riddle which has escaped satisfactory explanation for centuries. Perception is what connects us to our surroundings, it is the link between the physical objects in the outer world and our innermost sensations, thoughts and feelings. As the main channel between the outside world and our inner selves, perception plays a critical role in the life of all living creatures.

Given its importance, it is not surprising that many people have dedicated great effort to understanding what perception might be and how it might work. Epistemological effort regarding perception has a long and fascinating history. Over 2300 years ago Aristotle wrote in his *Metaphysics*:

All men by nature desire to know. An indication of this is the delight we take in our senses; for even apart from their usefulness they are loved for themselves; and above all others the sense of sight. For not only with a view to action, but even when we are not going to do anything, we prefer seeing (one might say) to everything else. The reason is that this, most of all the senses, makes us know and brings to light many differences between things.

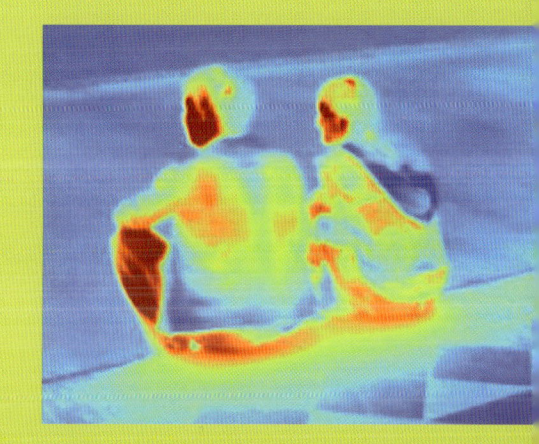

Aristotle was one of several great philosophers who taught that each of us has direct and immediate acquaintance with our own personal sensations; thus errors in our understanding are not due to sensation, but rather to the attempts we later make to reflect and understand. From the time of Aristotle, if not earlier, we find that perception is placed at the heart of what it means to be human. Understanding perception is a very serious matter indeed, as its ramifications reach into almost all fields of human endeavour.

Despite the many observations and reflections which have been made over the centuries perception remains a much misunderstood concept to this day. What we perceive, what we can perceive, and how we can perceive are topics which have not often been addressed satisfactorily. Modern life, governed as it is by imagery and brand, whether in the realm of the private or the public, the spiritual or the commercial, often neglects the question of what exactly our perception is telling us. An infinity of examples can be found in a multitude of fields in which messages are communicated without reference to perception, with perception considered simply a vehicle for thought and emotion. Perception usually remains unquestioned, unexplored, neglected; it simply is. The sensuousness of a fragrance, the spiciness of a food, the eroticism of a beautiful body: all are prompted by images or sounds which stimulate thought and emotion almost directly, largely bypassing detailed inspection. Perception is usually taken for granted. It does its job. It is.

Despite the incredible work performed by our perceptual and cognitive systems, they are nevertheless often either underused or overused, and can be misinformed or even deceived. Avoiding perceptual deception has been suggested by some to be the single most important problem in life. In popular culture a number of common sayings such as "seeing is believing", "trust your own eyes" and "so close I can taste it" are evidence of the general acceptance of the fundamental role of sensory perception. What little we know about the world around us is mostly obtained through our sensory abilities, therefore some understanding of their function and limits, their merits and demerits, is fundamental. This seems reasonable, even obvious. This is, however, not usually the case.

It is perhaps precisely because of its central role that perception is often neglected, forgotten, put to one side. Debating the merits or demerits of perception can produce doubts and worries, and can lead to uncomfortable questions about our relationship with the world around us. If perception is put under scrutiny, then so is the world which perception has constructed, and so is our understanding of ourselves. Perception can therefore be a worrying topic, something like an unpleasant relative who cannot be renounced, but who is avoided whenever possible. Perception is the kind of topic which should be avoided at the dinner table.

Can we then simply take perception as a given? Is it possible to delve deeper into the structure and meaning of the universe without delving deeper into the structure and meaning of our own perception? The answer to such a question is probably no. It is at least likely to be no. This book exists because I strongly believe that it is no.

Thinking of our lives in the 21st century and the things which we produce, one cannot help noticing that the operational limits of much of what we do coincides with the limits of human perception. In a world in which it appears that anything is now possible, the limitations of many forms of human endeavour are not those dictated by the physical world, but rather those dictated by our own perception and thought. Designers across the world realise that people are now at the centre of their work and that human perception and thought define the new frontier. Where material, form or manufacture once dominated their work, perception, cognition and emotion are the new black. Like successful design, successful life in the 21st century is a human centred affair. In the great machinery of our 21st century world, perception is a lynch pin.

The aim of this book is to propose an approach to the ancient riddle of perception. The great philosophers from Aristotle to Sartre have debated the nature and limits of perception using the tools of language and logic, and science has constrained perception to the statistical outcomes of focused

page 14: Tighter fitting or minimal layers of clothing reveal more body heat

page 15: The heat from these two people meets on the surface of the pavement

opposite: Heat is lost through the windows of a building

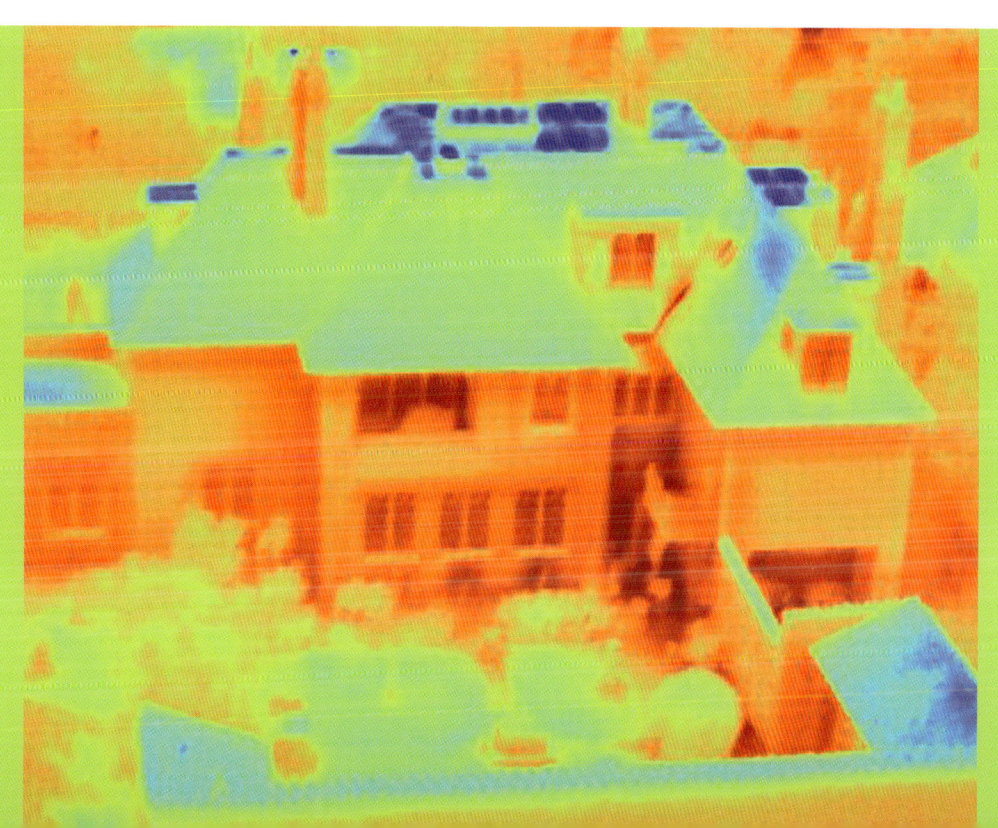

right: A thermal image can indicate which parts of a house absorb or reflect heat and sunlight

laboratory experiments, but this book wishes to raise a voice in favour of a more experiential, interactive and design-based approach. By means of graphical imagery this book attempts to illustrate how unusual and alien our world can seem when glimpsed through eyes different from our own biological equipment. New technologies come to market every day which permit radical new insights into the world around us, allowing us to explore nature with eyes and ears different from our own. Like children on Christmas eve, we now find ourselves surrounded by a dazzling array of new 21st century toys which no philosopher or scientist of the past could have dreamed of. Rather than speculate on what perception tells us about the external world, the approach proposed here is to experience. Rather than rely on traditional understanding, the recommendation is to expand the envelope. Like the psychedelic explorations of the 1960s, new technologies are now opening a previously unknown world of perception for our benefit.

This book is intended as a primer for those who are interested in exploring what the world may actually consist of, as opposed to what it is normally assumed to be. The idea is simple: use new technologies to perceive the world as it has never before been perceived. Human exploration is entering a new phase, one in which it is not necessary to travel to the antipodes to discover alien new worlds. The weird and wonderful can be found in our own neighbourhood, in our own backyard, in our own home and maybe even within ourselves. To borrow a term popularised by the great philosopher Immanuel Kant, this book proposes the benefits of using new perception enhancing technologies to explore the "things in themselves".

For this book, a medium was required for illustrating the possibilities which 21st century technologies hold for expanding our understanding of the world around us. Among the current spectrum of technologies it is immediately obvious that several are already in regular use by various professionals and that they are already changing the way we view our world. Among the technologies which produce

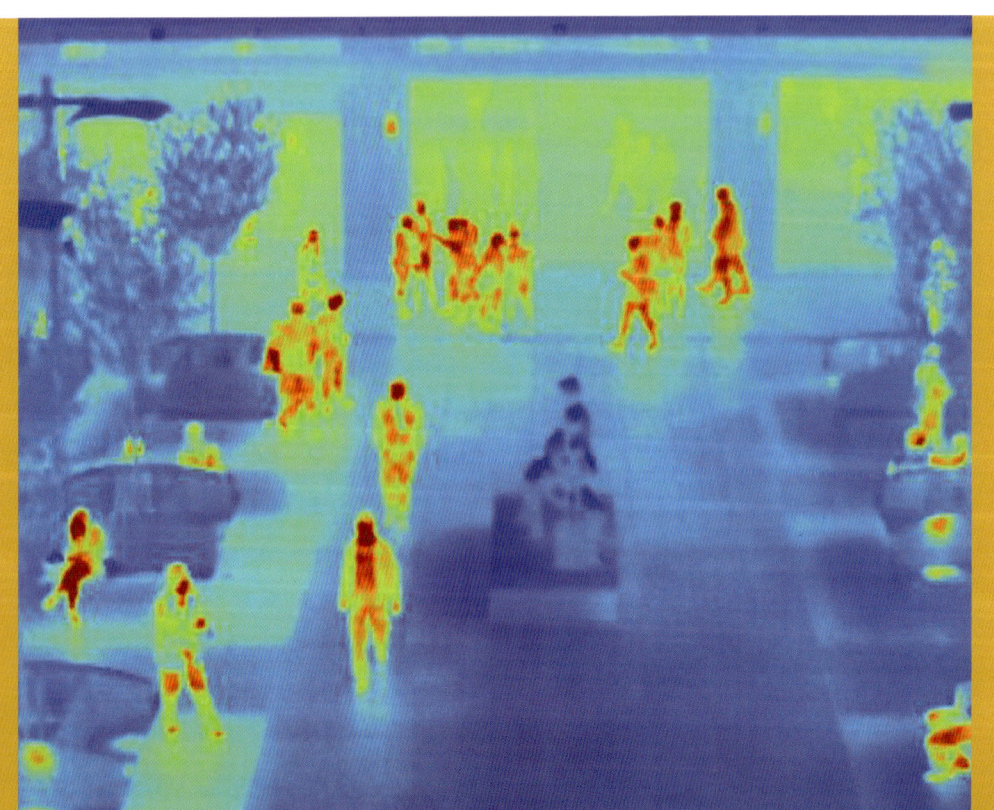

left: Thermal vision quickly separates people from their environment at this shopping centre, revealing geometric patterns of motion

opposite: As an aeroplane flies overhead, the hot engines are clearly visible against the lower temperature of the sky

page 20: A close-up portrait reveals the different thermal textures of skin and hair

page 21: The clothing and accessories of this woman exhibit varying amounts of heat based on their closeness to the body

outputs appropriate to reproduce on paper, thermal imaging was chosen because of the many insights which arise from seeing heat. Like the great universal fire of the Stoics, the world around us is a cauldron in which objects are characterised by heat, where heat is unequally distributed, and where heat moves from place to place. Thermal images thus provide a new and exciting window into the world of the "things in themselves", revealing phenomena which we do not normally see. This book contains a collection of thermal images which I hope will provoke, lead to opinion and entice some of you to give perception enhancing technologies a go. It is my heartfelt hope that the designers of the future will never start a project without first taking a good look around through perception enhancing eyes.

20

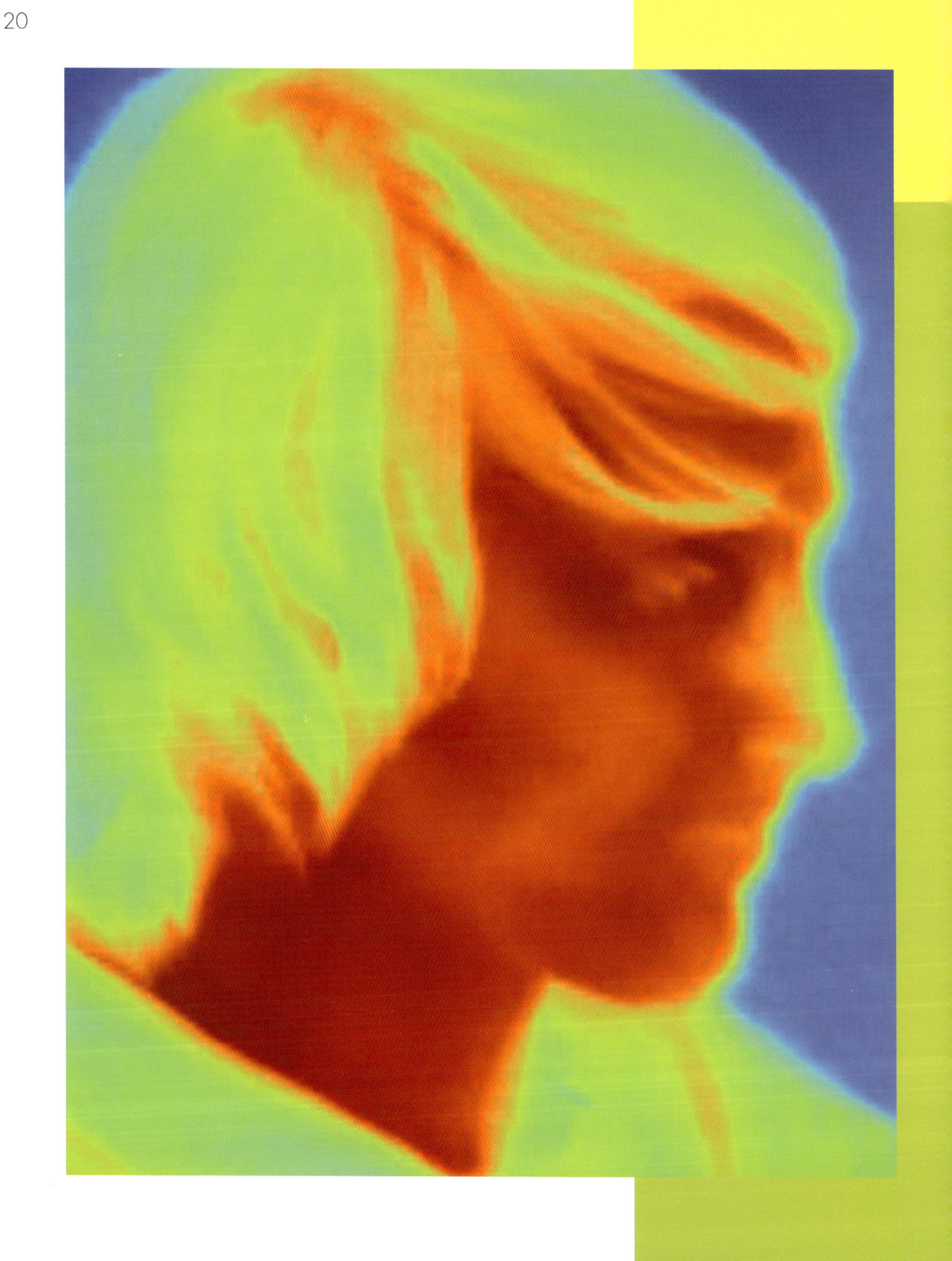

WHAT IS PERCEPTION?

The question of how we perceive the world is among the most ancient of the intellectual riddles. Along with questions about fire and tools, our ancestors must have spent time reflecting upon how they perceived the world around them. The birth of language, not later than the arrival of humans in Australia approximately 50,000 years ago, must have led to the birth of debate about the nature of perception. In Western society, the tradition of debating and asking questions has developed into the discipline which we call philosophy, which in its most advanced form consists of structured analysis of problems which are central to the human condition. The two areas of philosophic endeavour which most closely address the subject matter of this book are metaphysics, which is the study of the types of things which exist, and epistemology, which deals with the basis and nature of knowledge. Philosophical developments have provided concepts and structures with which to consider the nature and role of perception, and so a brief review of philosophical thought on the subject provides a background against which to judge the contents of this book.

Perception in the Classical World

Evidence of Western philosophical thought about perception can be traced back to at least the 7th century BC when men from around the Mediterranean developed theories about how knowledge is, or is not, derived from the information provided by the senses. An early contribution was provided by Alcmaeon of Croton (c. 540-500 BC) who suggested that the brain was the seat of the soul, and that the sense organs were connected to it by means of channels. He was one of the first philosophers, perhaps the first philosopher, to conclude that perception is the source of all knowledge. His analysis of the sources of knowledge has lead many to attribute to him the foundation of what we today call epistemology.

GLOBAL WARMING

Though thermal cameras cannot directly measure the effects of global warming, it is nevertheless hard to ignore the problem when you view the world through thermal eyes. Detecting heat directly through thermal vision provides a worrisome image of the modern world. A glance around suggests a landscape of cool natural materials, particularly biological material such as plants, against which we have constructed a world of temperature accumulating objects. A cool blue tree silhouetted against a hot yellow building cannot fail to raise questions about how people have modified the environment. Can enhanced perception help us to understand how to save our planet?

above: A cast iron sun-warmed post box stands out against the background vegetation

below: The surrounding tree branches display different thermal properties to the neighbouring garden shed. The leaves are much cooler than the wooden shed, which has been heated by the sun

One of Alcmaeon's contemporaries, Heraclitus of Ephesus (c. 535-475 BC), taught instead that sensory perception is necessary to gain knowledge but not sufficient. According to Heraclitus, people must also possess the ability to interpret things. He championed the idea that sensory perception is reasonably reliable but that the conclusions which we draw from it might not be. Typical of Heraclitus' thinking was the phrase "eyes and ears are bad witnesses for those who have souls that do not understand the language".

The physical mechanisms underlying sensory perception were instead extensively studied by Anaxagoras of Clazomenae (c. 500-428 BC) who proposed a theory which was based on differences. Situations such as a warm hand touching a cold stone suggested to him that sensation was the result of the interaction of things of dissimilar nature. Anaxagoras proposed that perception was the ability of the soul to note the changes which occur in one or the other of the dissimilar things when their materials come into contact.

A diametrically opposite conclusion to the problem was drawn by Empedocles (c. 490-430 BC) who believed instead that perception was the result of the interaction of similar things. He taught that the bodies of animals and of humans incorporate pores which receive material of similar type from external objects. According to Empedocles, perception was not passive, but instead an active process of controlling the pores which link us to the external world. Following through with this logic he also emphasised the relatively limited capabilities of human perception, suggesting that our biological equipment provides us with only poor and inaccurate versions of the truth. Like Heraclitus before him, Empedocles taught that a philosopher must apply the thinking processes of the soul to the task of understanding the world. Perception, according to Empedocles, was not sufficient to achieve truth.

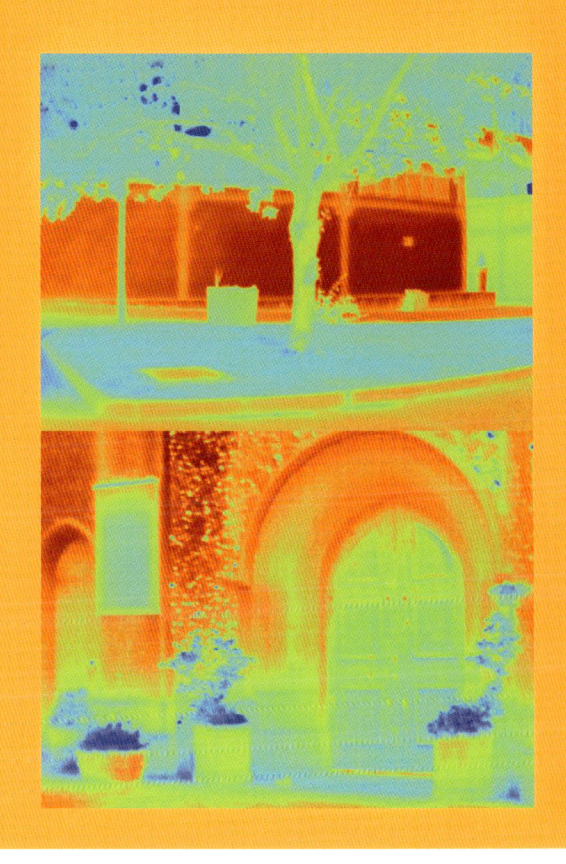

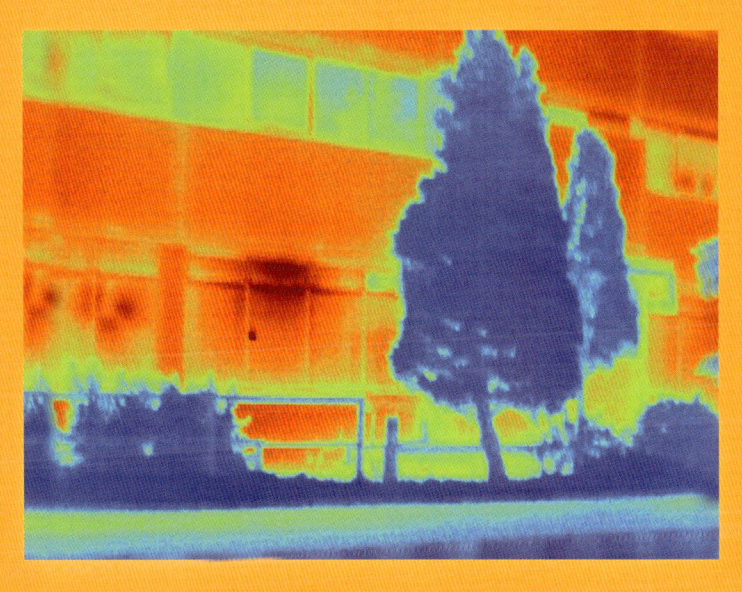

above left: Cool grass and a shady tree form a tranquil corner by this building

left: Lichen, moss and plants are thermally cooler than the small stone church they grow on

above: A young tree and surrounding vegetation display much cooler temperatures than the man-made office building directly behind them

A conceptual milestone in Western philosophy was reached with the work of Protagoras of Abdera (c. 480-410 BC) who noted that different people have different perceptual abilities, leading to the logical conclusion that perception could not be absolute. Taking this one step further, if perception was not absolute, then neither could truth be absolute, since it depended on perception for its foundations. The logical impossibility of achieving truth by means of only perception is a concept which has important ramifications for how we conduct philosophy and science. These insights into the relativistic nature of the human condition are neatly summarised by Protagoras' saying "man is the measure of all things" which is still well known today.

Democritus of Abdera (c. 460-362 BC) developed a revolutionary model of the universe based on elementary particles called atoms. According to Democritus the physical universe consisted of an empty void and of an infinite quantity of indivisible and eternal atoms. Democritus' theory of perception was that atoms travelled from objects through the air and interacted with similar atoms contained in our sensory systems, creating an image of moving atoms which flowed to the soul. Like others before him, he firmly believed that all knowledge is obtained from sensory perception. Also, as with several of his predecessors, he believed that individual differences in sensory abilities suggest that we have no way of knowing whether our own perceptions correctly represent the world around us, or whether they are the same of those of others. According to Democritus:

We know nothing for certain, but only the changes produced in our body by the forces that impinge upon it.

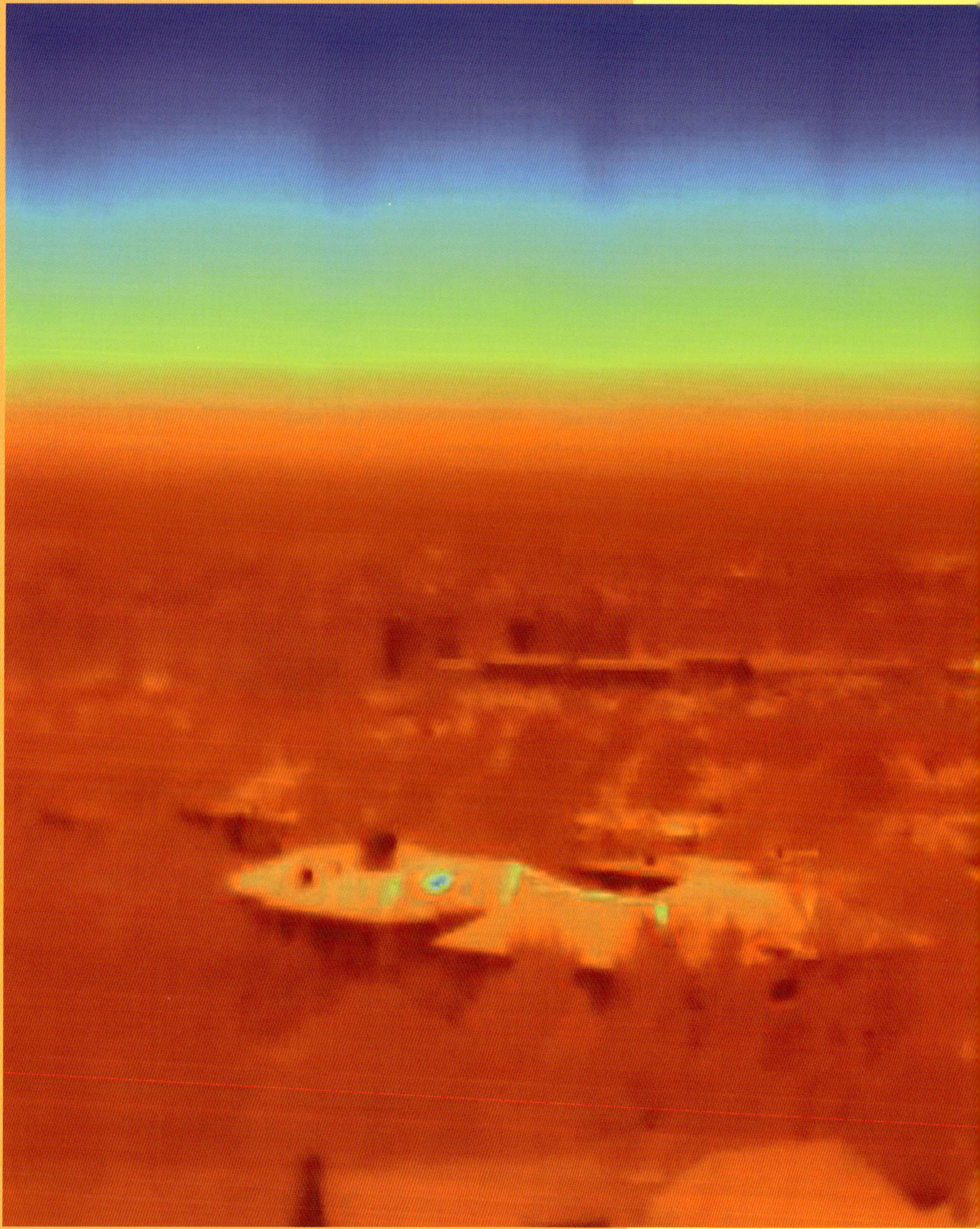

25

A evening view over the city, showing the different levels of heat in the atmosphere. The warmest temperatures surround the earth, cooling up through the atmosphere to the colder regions out towards space.

The famous Plato (c. 424-347) also asked many epistemological questions about the nature of knowledge. He was convinced that differences existed between our sensory experience and physical reality. His theory of the forms was intended as a framework for understanding things in the world, and typical of his thinking is the allegory of the cave in *The Republic*, where men grow up seeing only shadows on a wall rather than the true forms of those who are outside the cave. According to Plato, trusting sensory information is like the prisoners of the cave trusting the shadows on the wall instead of the true shapes of the people who caused them. His emphasis on the need to think abstractly in order to make sense of the world is exemplified by the phrase which he had Socrates utter in the *Theaetetus*:

The simple sensations which reach the soul through the body are given at birth to men and animals by nature, but their reflections on the being and use of them are slowly and hardly gained, if they are ever gained, by education and long experience.

Plato's one time student and eventual rival Aristotle (c. 384-322 BC) devoted great attention to the study of perception, considering it to be the capacity which distinguishes animals from plants.

ELECTRICITY

The discovery of electricity is one of the unsung triumphs of the human race, close to the discovery of fire in its significance, but often taken for granted in today's world of push-button devices. As a form of energy in motion, electricity cannot be prevented from creating heat in most appliances due to resistance in the materials along which it travels. Where electricity flows, heat is sure to be present. Thermal eyes can thus speedily divide the environment around us into what is electrical and what is not based on what exhibits heat and what does not.

right: A mobile telephone battery and charger heat up as electricity flows through them. The wire is not as hot as it is heavily insulated

opposite above left: Underground electrical wiring. The wires heat to different temperatures as they carry different amounts of current

opposite below left: People and electrical devices stand out from their environment at this road junction

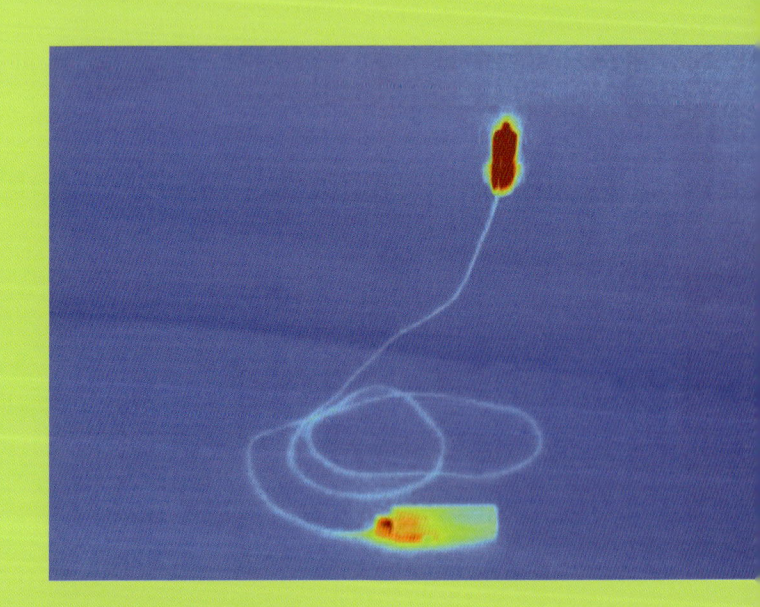

Typical of his views is this section from *De Anima*:

That perceiving and practical thinking are not identical is therefore obvious; for the former is universal in the animal world, the latter is found in only a small division of it. Further, speculative thinking is also distinct from perceiving, I mean that in which we find rightness and wrongness; rightness in prudence, knowledge, true opinion, wrongness in their opposites; for perception of the special objects of sense is always free from error, and is found in all animals, while it is possible to think falsely as well as truly, and thought is found only where there is discourse of reason as well as sensibility.

In several of his works, Aristotle claimed that every animal had at least the sense of touch, and some animals had other senses as well. In *De Anima* he further proposed what is considered to be the first formal classification of the five basic senses of sight, hearing, smell, taste and touch, a system which has remained with us to this day. Aristotle's efforts to understand perception were in part due to his early observations that perception appeared to be the ability by which animals nourished themselves and navigated the world. Thus, for Aristotle, perception merited study because there could not be life without it.

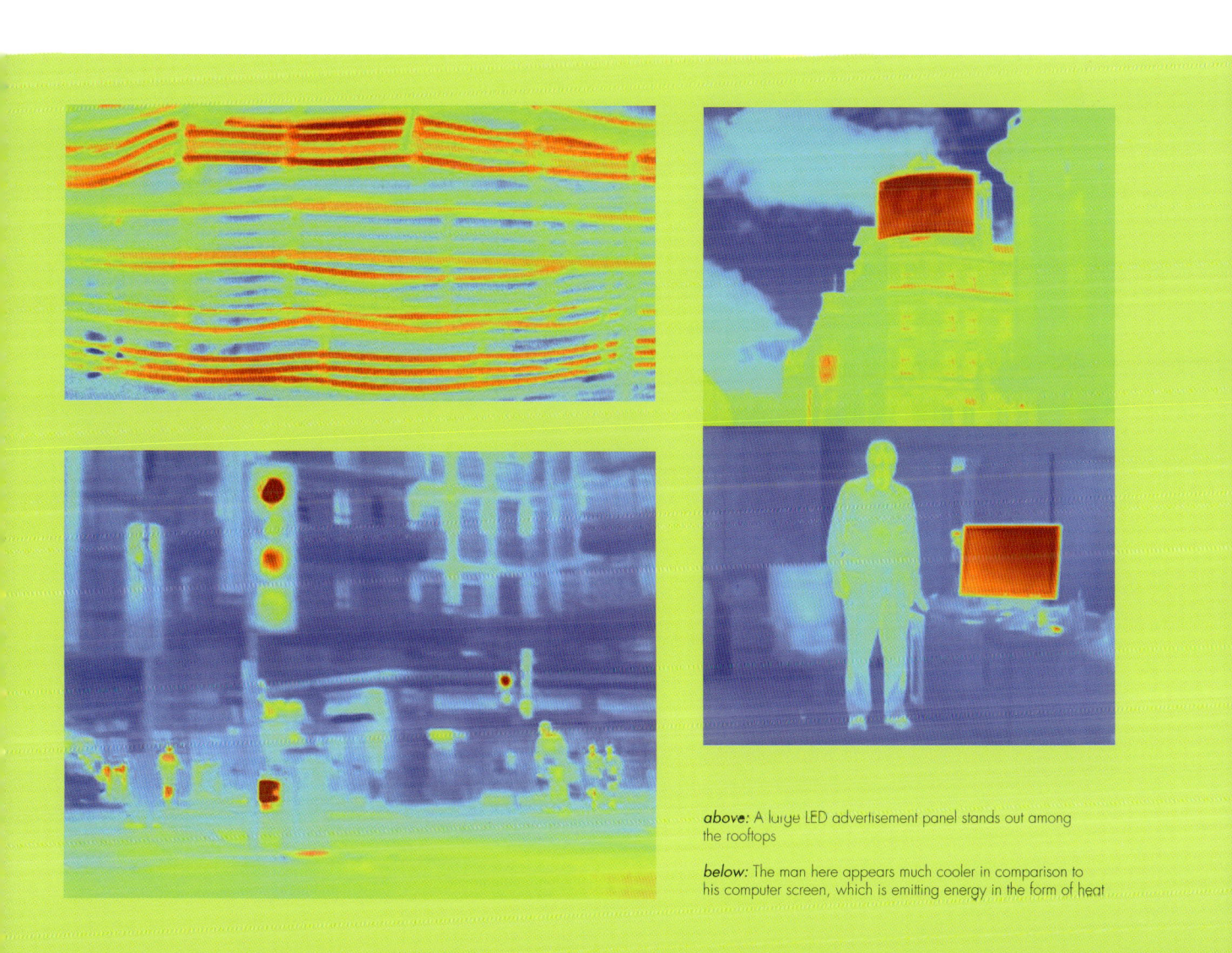

above: A large LED advertisement panel stands out among the rooftops

below: The man here appears much cooler in comparison to his computer screen, which is emitting energy in the form of heat

HEAT AND FIRE

Since the dawn of time people have been fascinated by fire. Fire and heat are fundamental to our lives and this dependence is dramatically brought to our attention by thermal eyes. Instead of detecting the presence of heat through the thermoreceptors of our skin, which requires proximity or even direct contact, heat can be made visible to the eye. Thus one sensory system is exchanged for another. This perception enhancement can transform the kitchen stove or the bathroom sink into fast changing, eerily coloured devices. The banal and everyday suddenly becomes unfamiliar and exotic.

above left: Hot water from a bathroom sink disappears down the drain pipe

above right: A home radiator shows that it is distributing heat evenly

right: Hot water from a bathroom sink tap spirals into the drain

The founding father of the Epicurean school of philosophy was another individual who taught extensively on the subject of perception. Like Democritus before him, Epicurus (341-270 BC) believed that atoms travelled out from objects and stamped themselves on our perceptual systems. Sensory images were the solidification of the atoms which arrived from external objects, and the basic senses were differentiated by the ease with which the external stimuli stamp themselves on the individual sensory organs. In his letter to Herodotus he articulated this mechanical view of perception by writing:

And we must indeed suppose that it is on the impingement of something from outside that we see and think of shapes.

For the Epicureans, the atoms which streamed off of external objects produced not just raw sensation but also subsequent thought. Thought was considered to be a form of image, similar to that of perception, only thinner and more fragile, produced by the bouncing and reflection of atoms inside of us. In the Epicurean world view, thoughts were manipulated by the soul by means of active powers of belief and preconception, bringing images into focus and leading to a gradual solidifying of the image into a concept. Writing several centuries later in *The Lives and Opinions of Eminent Philosophers*, Diogenes Laertius (3rd century AD) summarises the writings of Epicurus:

By preconception, the Epicureans mean a sort of comprehension as it were, or right opinion, or notion, or general idea which exists in us; or, in other words, the recollection of an external object often perceived anteriorly. Such for instance, is this idea:

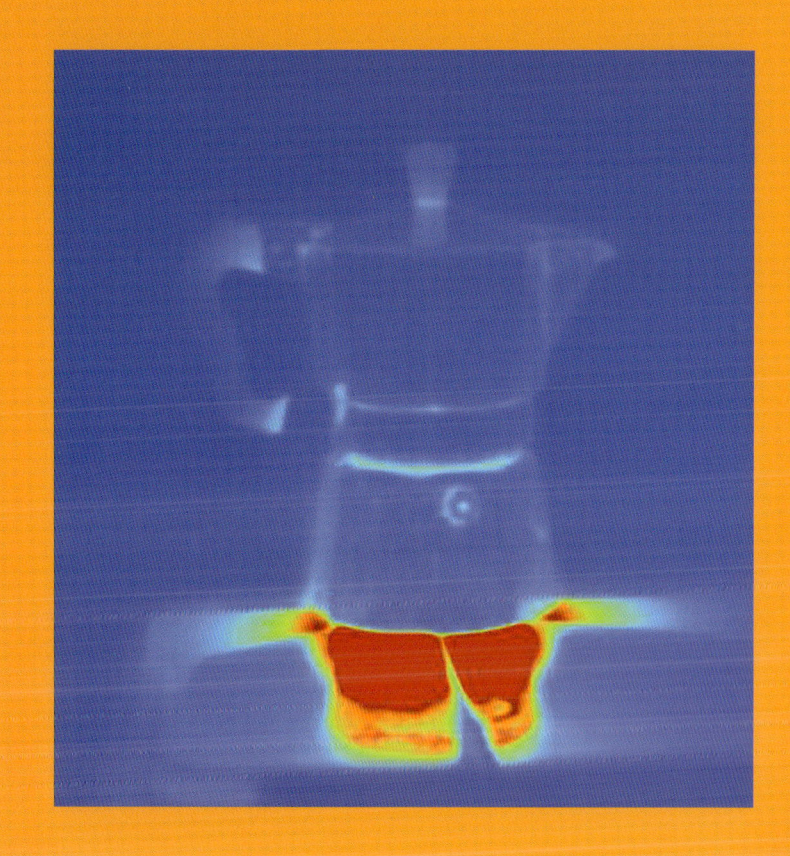

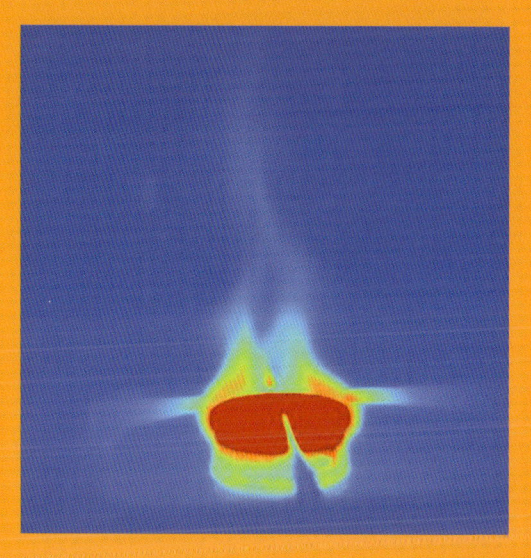

left: The flames from the aluminium base of a kitchen stove burner just begin to heat a coffee machine

above: A column of hot air rises from the flame of a kitchen stove

"Man is a being of such and such a nature." At the same moment that we utter the word man, we conceive the figure of a man, in virtue of a preconception which we owe to the preceding operations of the senses. Therefore, the first notion which each word awakens in us is a correct one; in fact, we could not seek for anything if we had not previously some notion of it. To enable us to affirm that what we see at a distance is a horse or an ox, we must have some preconception in our minds which makes us acquainted with the form of a horse and an ox. We could not give names to things, if we had not a preliminary notion of what the things were.

As with Epicureans, the Stoic school of philosophy (founded by Zeno of Citium in the 3rd century BC) also developed a model of perception whereby external objects impressed themselves on the soul. More so than the Epicureans, the Stoics emphasised the active component of the process, developing explanations for how differences in people's background and character led to differences in perception. A voluntary act of the soul was required in order to identify and understand the objects that were being perceived. What must be added is what the Stoics called "assent", and the wise man was considered he who was careful in allowing "assent" to sensory impressions.

Despite the active role assigned by the Stoics to the soul they developed no specific theory to describe the act of thinking. For example, hallucinations and dreams were considered to be aberrations caused by poor functioning of the sensory organs. The extreme centrality of pure sensation in the Stoic world view is well expressed by the words of Diogenes Laertius in *Lives and Opinions of Eminent Philosophers*:

Also, all notions arise from the senses by means of confrontation, analogy, similarity and combination, with some contributions from reasoning too.

One final contribution of the Stoics to the topic of human perception was the line of enquiry elaborated by Hierocles (c. 200 AD) that self-perception is the most basic and important ability of all animals. His well known phrase "all perception is self-perception" summarises the Stoic view that all animals perceive themselves from the moment of their birth and that their very first perception is a self-perception. In the eyes of the Stoics, therefore, sensory perception was not a relationship only between ourselves and the outside world, it was also a relationship between our own mind and body.

Perception During the Renaissance and the Enlightenment

During the Middle Ages the theories of the philosophers of the ancient world continued to be developed and elaborated, with many works of Aristotle finding particular favour after their reintroduction into the West. But with the passing of time, the world view of the ancients began to creak and groan under the

PEOPLE AND ANIMALS

As highlighted by Hollywood movies such as *Terminator* and documentaries such as *The Cove*, warm blooded creatures are immense sources of heat. By having thermoregulatory systems which maintain a near constant core temperature, they stand out like the proverbial sore thumb against the cool background of the environment. Warm blooded creatures cannot hide from perception enhancing thermal eyes; they leave a thermal trail behind them wherever they go.

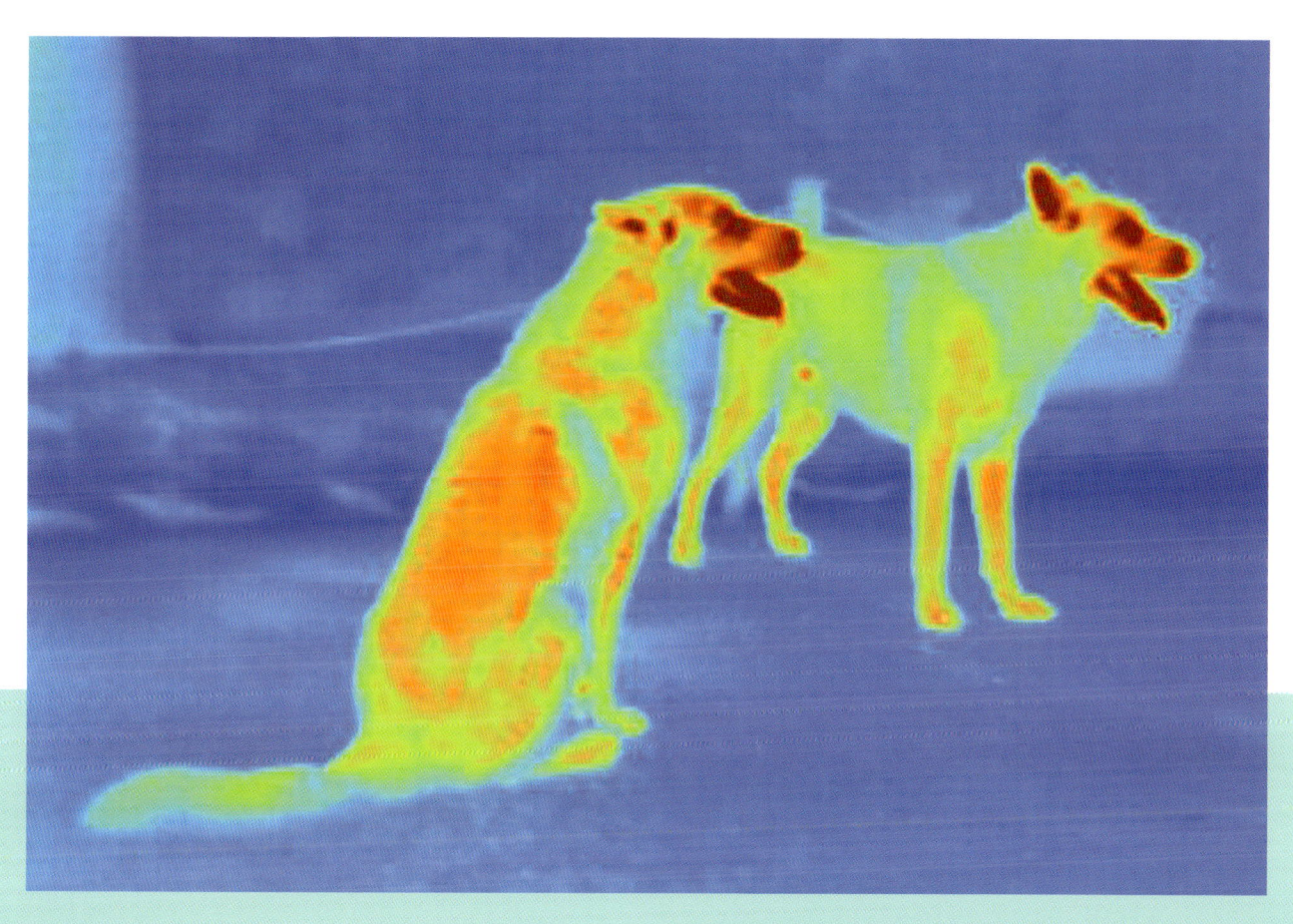

opposite: Groups of people stand out from the cool surroundings of the underpass as they exit an underground station. Note the different types of clothing: some people are wearing short sleeves, others long sleeves; some people are wearing trousers and others shorts or skirts

above: A pair of dogs dissipate heat through their mouths as they pant in the warm weather

below left: Two swans drift gracefully on a stream on a warm summer's day. Thermal vision emphasises the nervous system by highlighting the great warmth which is found at the head and neck, near the brain

below right: Grazing sheep show off the hot spots on their bodies, specifically their heads and udders which are not covered by their thick woollen fleece

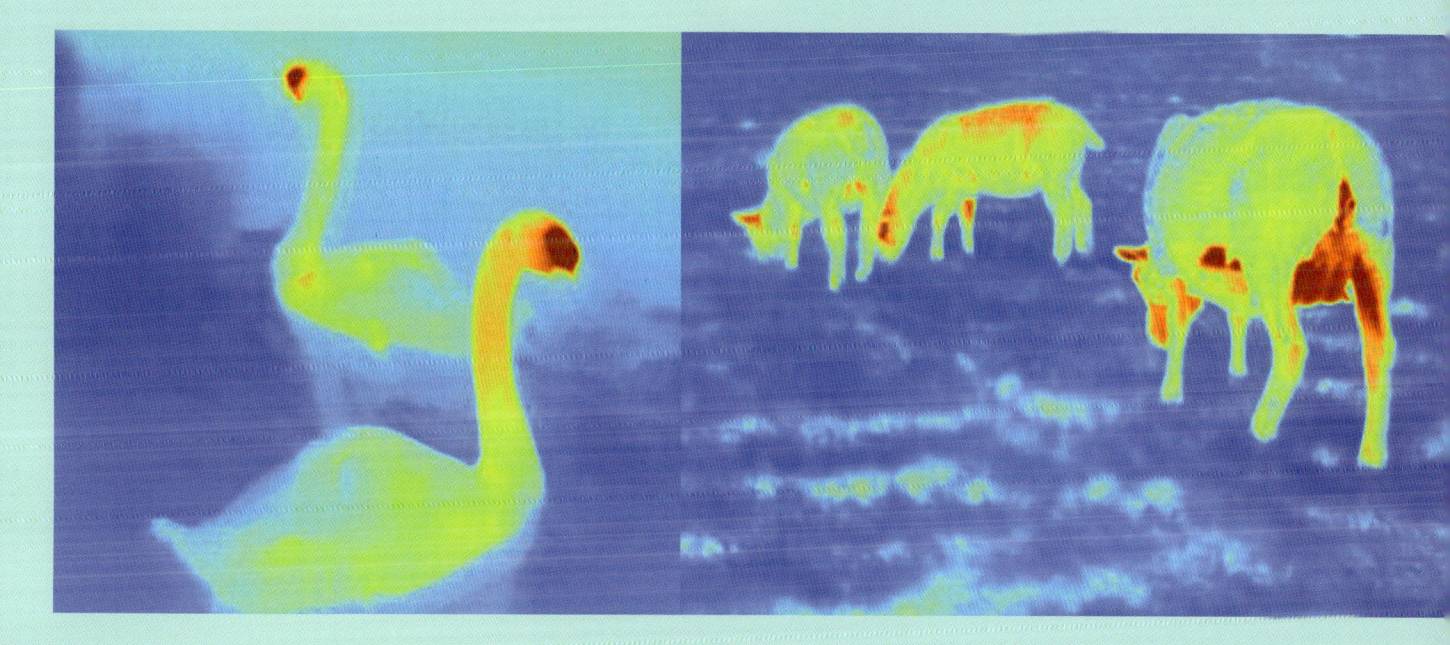

MOTION

Looking at the world through thermal eyes quickly leads to the conclusion that motion and heat are inextricably linked. Materials have properties which lead to friction, and friction produces heat. Quite simply, nothing moves without producing heat. Vehicles and transport systems are particularly fascinating when viewed through thermal eyes because their motion-generating components become hot; something which can sometimes be noted through normal vision, but which is dramatically enhanced in a thermal world. The engines, motors and other moving parts are easy to spot in these images.

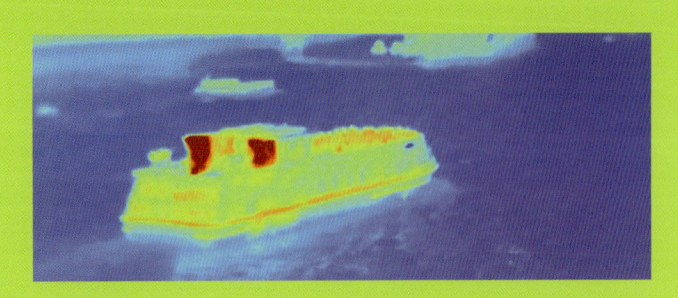

below left: The exhaust funnels of this river cruise boat can be clearly seen as they expel hot air

below: Heat indicates the location of the engines of an overhead aeroplane

opposite, below: Heat quickly reveals the location of a car's engine and a lamp's light bulb

weight of new scientific discoveries. The Renaissance philosophers of the 15th and 16th centuries began to break with tradition on many fronts, producing radical new views about problems ranging from the orbits of the planets to the functioning of human perception. For example, Bernardino Telesio (1509-1588) was one of a group of philosophers who left university in order to develop ideas which transcended Aristotelian tradition. In his work *On the Nature of Things according to their Own Principles*, he rejected the ancient concept of sense organs, replacing it with a more mechanical process in which external impressions were transmitted through nerves to the brain:

Sense perception can only be the perception of the activities of things and impulses in the air, and can only consist of the perception of the spirit's own passions, transformations and movements, particularly the latter. Indeed, the spirit perceives them because he perceives that it is affected by them, that it is being changed and moved.

Of the philosophers of the later Renaissance the individual who left us the greatest legacy in the field of human perception is René Descartes (1596-1650), who many consider to be the father of modern philosophy. In his best known work, *Meditations on First Philosophy*, he reflected at length on the relationship between body and mind, considering what people can know with any degree of certainty starting from the evidence provided by the senses. Descartes put sensory perception at the centre of his epistemology, as illustrated by a well known passage from book one of *Meditations on First Philosophy*:

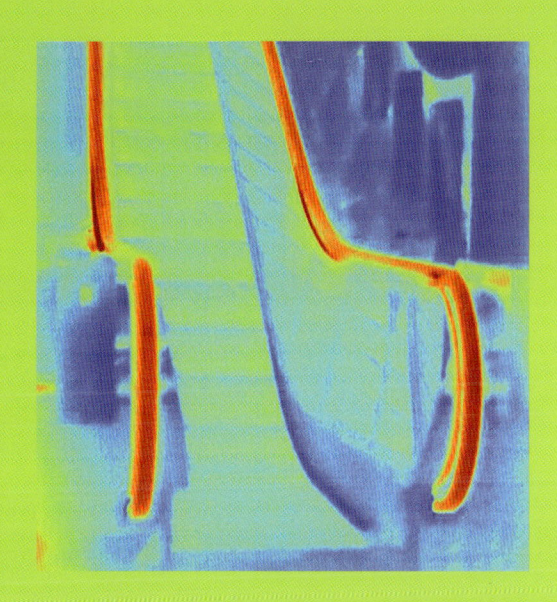

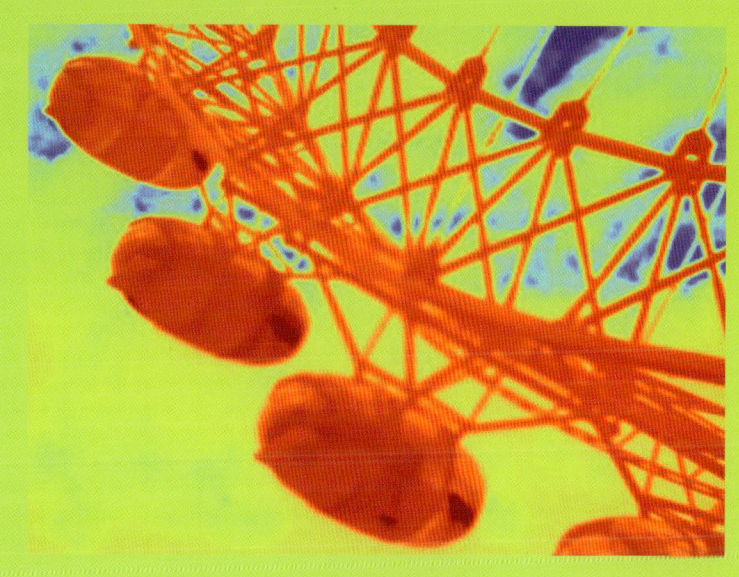

above left: Friction causes the handrails of this escalator to heat up as it moves

above right: The darker sections of red on these pods of the London Eye reveal the location of the gears and motors inside

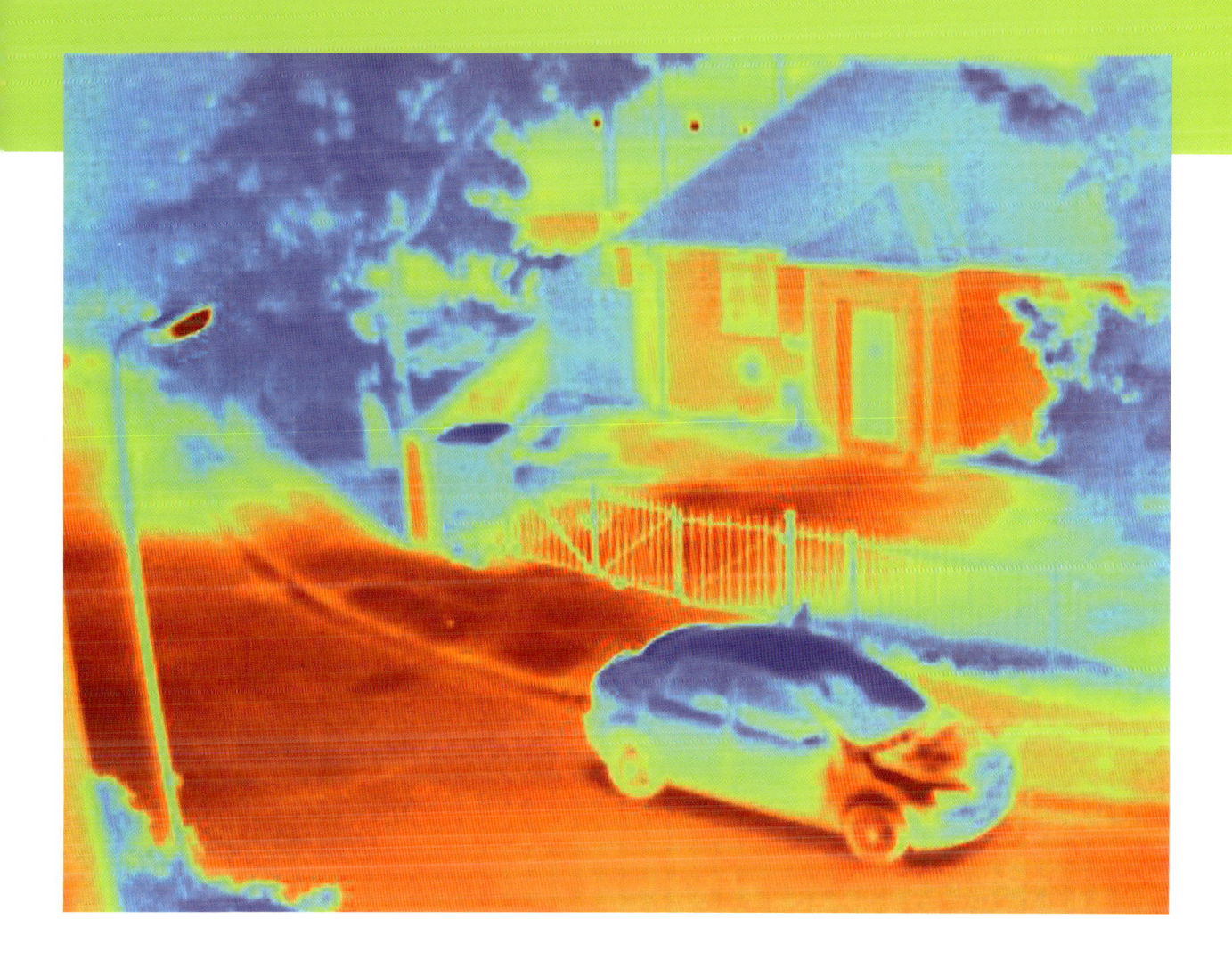

But it may be said, perhaps, that, although the senses occasionally mislead us respecting minute objects, and such as are so far removed from us as to be beyond the reach of close observation, there are yet many other of their presentations, of the truth of which it is manifestly impossible to doubt; as for example, that I am in this place, seated by the fire, clothed in a winter dressing gown, that I hold in my hands this piece of paper, with other intimations of the same nature. But how could I deny that I possess these hands and this body, and withal escape being classed with persons in a state of insanity, whose brains are so disordered and clouded by dark bilious vapours as to cause them pertinaciously to assert that they are monarchs when they are in the greatest poverty; or clothed in gold and purple when destitute of any covering; or that their head is made of clay, their body of glass, or that they are gourds? I should certainly be not less insane than they, were I to regulate my procedure according to examples so extravagant.

In his book, Descartes also dispelled with the scepticism of his contemporaries regarding the deficiencies of the senses, establishing a strong positive case for the centrality of sensory perception in human endeavour. In another well known passage from book one he established his positivist outlook:

SUNLIGHT

Thermal eyes provide a striking change to the way we see our world by introducing the idea of thermal memory. The naked eye can see where light is instantaneously emitted, absorbed or reflected, but thermal eyes reveal not just where light and energy are now, but also where they have been previously. Materials accumulate and store energy, creating a ghostly trail which lingers long after the sun has become hidden behind the clouds or set for the evening.

below left: A church is warmed by the afternoon sun. The stone is much cooler where the shadow lies across the lower right half of the building

below: Grass, stone and steel create a spectacular colour display when viewing London's Tower Bridge from the north

opposite top left: A tree in a field casts a thermal shadow, indicating the direction of the sun

opposite top right: Sunshine warms one side of a cast iron telephone box

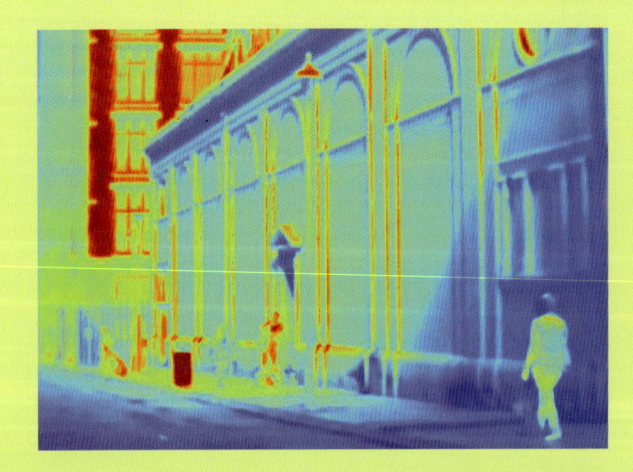

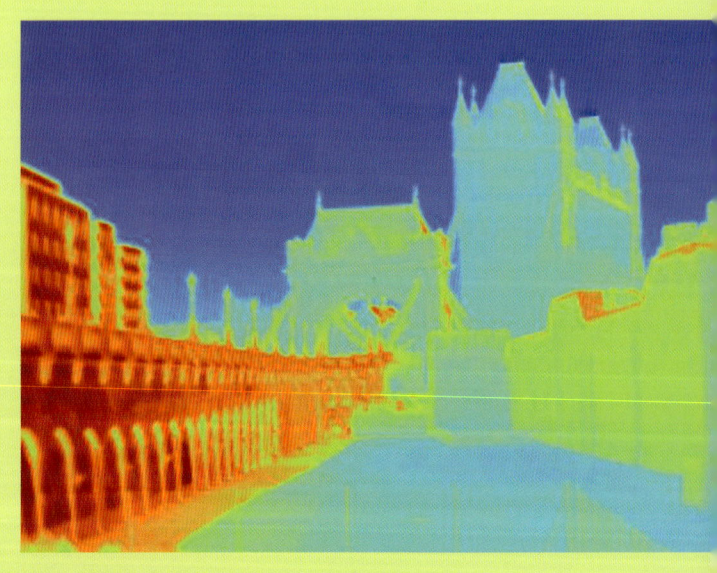

Let us suppose, then, that we are dreaming, and that all these particulars – namely, the opening of the eyes, the motion of the head, the forth putting of the hands – are merely illusions; and even that we really possess neither an entire body nor hands such as we see. Nevertheless it must be admitted at least that the objects which appear to us in sleep are, as it were, painted representations which could not have been formed unless in the likeness of realities; and, therefore, that those general objects, at all events, namely, eyes, a head, hands, and an entire body, are not simply imaginary, but really existent. For, in truth, painters themselves, even when they study to represent sirens and satyrs by forms the most fantastic and extraordinary, cannot bestow upon them natures absolutely new, but can only make a certain medley of the members of different animals; or if they chance to imagine something so novel that nothing at all similar has ever been seen before, and such as is, therefore, purely fictitious and absolutely false, it is at least certain that the colours of which this is composed are real. And on the same principle, although these general objects, viz. a body, eyes, a head, hands, and the like, be imaginary, we are nevertheless absolutely necessitated to admit the reality at least of some other objects still more simple and universal than these, of which, just as of certain real colours, all those images of things, whether true and real, or false and fantastic, that are found in our consciousness, are formed.

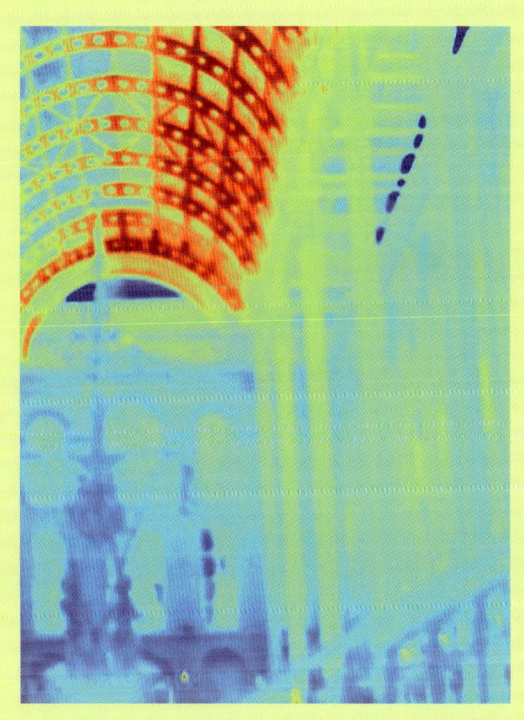

above: The sun warms the steel roof structure of this shopping arcade

left: A row of motorcycles and scooters is warmed by the afternoon sun

Having established his approach, Descartes moved on in book two to add his greatest contribution to the philosophy of perception:

Let us pass, then, to the attributes of the soul. The first mentioned were the powers of nutrition and walking; but, if it be true that I have no body, it is true likewise that I am capable neither of walking nor of being nourished. Perception is another attribute of the soul; but perception too is impossible without the body; besides, I have frequently, during sleep, believed that I perceived objects which I afterward observed I did not in reality perceive. Thinking is another attribute of the soul; and here I discover what properly belongs to myself. This alone is inseparable from me. I am, I exist: this is certain; but how often? As often as I think; for perhaps it would even happen, if I should wholly cease to think, that I should at the same time altogether cease to be. I now admit nothing that is not necessarily true. I am therefore, precisely speaking, only a thinking thing, that is, a mind, understanding, or reason, terms whose signification was before unknown to me. I am, however, a real thing, and really existent; but what thing? The answer was, a thinking thing.

This insight has acted as a magnet for enquiry and debate ever since, presenting a problem which

CLOTHING

Thermal vision gives us the ability to see the relationship between body and clothes. Thermal exchange between people and their surroundings is heavily moderated by both the thickness of clothing and the type of material from which garments are fashioned. Thermal vision quickly differentiates between thick cloth or thin, and can help us to understand the effects of different materials such as leather, silk, cotton and so on. What will fashion designers do once given the ability to see what their clients feel?

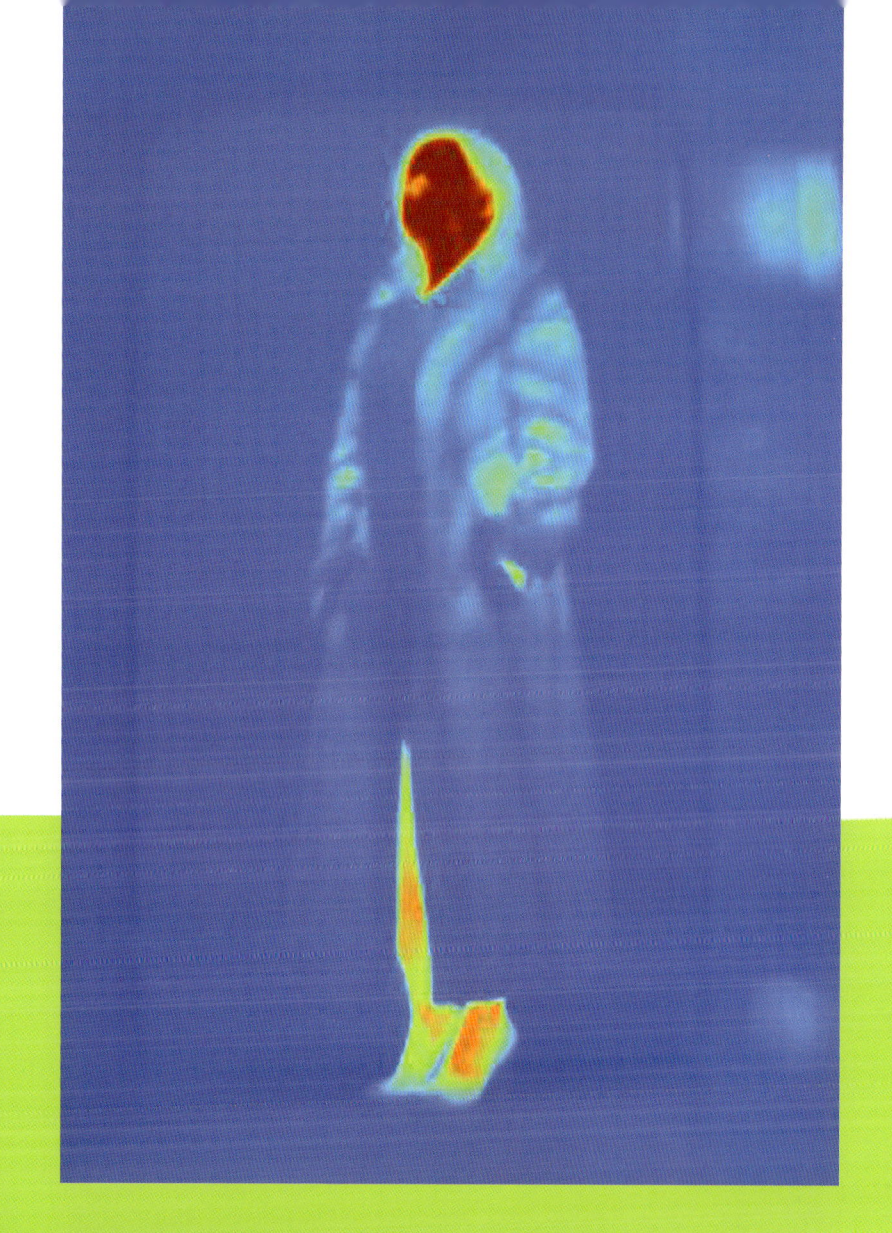

right: This long coat appears cool on the outside but keeps body heat trapped beneath it in cold weather

opposite and below: A model poses in a series of different outfits to illustrate the principle that different materials have different thermal properties

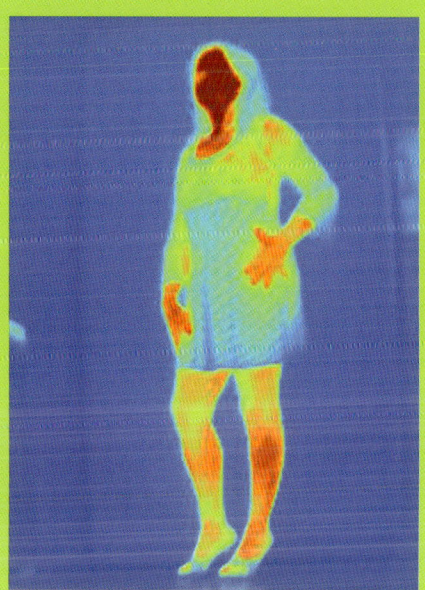

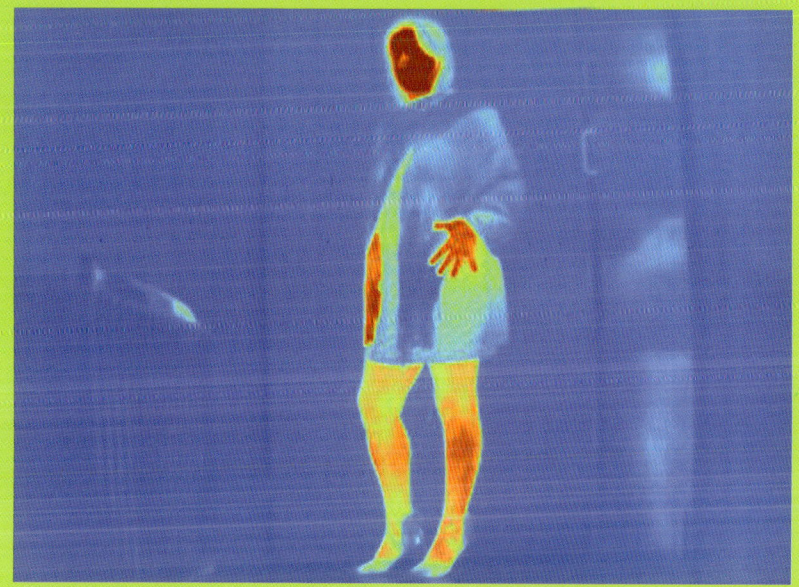

LIGHTING

Where there is light there is energy and where there is energy there is life. Even though this information is generally available, many people are still surprised to learn that the average street light or home lamp emits less than 3 percent of the inputted energy in the form of light, with nearly all the rest ending up as heat. To thermal eyes, this heat loss reveals a strange night time world of alien flowers, where natural colour gives way to intensity, and reflected light is replaced by the glow of the primary heat sources. The ability to see temperatures reveals a new geometry and a new perspective, with the eye drawn in odd and unexpected directions in response to a thermal landscape which we struggle to recognise as our own.

appears to defy adequate solution. Descartes' observation drove a wedge between body and mind, or, if preferred, between the physical and the subjective worlds. More so than anyone before him, Descartes put his finger squarely on the central problem of much of Western philosophy and science: the reconciliation of the external with the internal, the "out there" with the "in here". Descartes' point of view is neatly summed up by a single sentence from his *Discourse on Method*:

Cogito ergo sum (I think, therefore I am)

Descartes' equated our thinking human mind with our eternal soul, thereby relegating sensory perception and the rest of our physical bodies to a temporary earthly supporting role. Gone were the days of the ancients when thinking was considered a form of perception. With Descartes began the era of the body-mind problem: the search for the causal link between the objective world around us and the subjective world inside us. After Descartes, philosophy faced the enormous new challenge of reconciling traditional knowledge about ourselves with the facts and figures which were being rapidly generated by the emerging new physical sciences. Logical theory was about to meet scientific experiment and ideas were about to be tested in what was to that point the greatest expansion of knowledge which the human race had ever achieved.

One of the philosophers whose work ignited the explosion of rational thought which we today call the enlightenment was Thomas Hobbes (1588-1679). Among his contributions to philosophical enquiry were

opposite above: The lights of a carousel

opposite below: Spotlights outside a row of shops and restaurants at night

below left: Ceiling lights appear to guide the way inside this underground station

below right: The bright letters of city centre advertising stand out in the evening

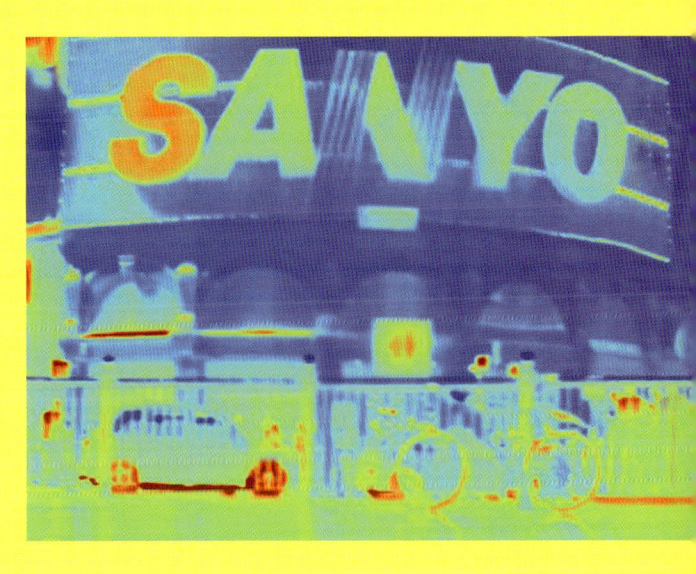

ideas for bridging the gap between the outer world of physics and the inner world of subjectivity. His purely materialist model of sensation and reason was based on a single underlying belief, summarised in his work *De Corpore*:

But the causes of universal things (of those, at least, that have any cause) are manifest of themselves, or (as they say commonly) known to nature; so that they need no method at all; for they have all but one universal cause, which is motion.

According to Hobbes, all perception and knowledge are the result of mechanical actions and principles, therefore nothing incorporeal can be truly known, only believed. Sensory organs respond to motion in the surroundings and communicate that motion through nerves to the brain. The mind works in a complex manner, but it too works according to the laws of physics and motion. In *Leviathan* he wrote:

The cause of sense is the external body, or object, which presseth the organ proper to each sense.

Another enlightenment philosopher, John Locke (1632-1704), further developed the empiricist world view that all knowledge comes from experience which is acquired through the senses. For Locke, abstract concepts like time, space, matter and numbers were the product of sensation followed by successive reflection. He denied that ideas and abilities were innate, describing the human mind instead as a *tabula rasa*, or "blank tablet", written on by experience. In *An Essay Concerning Human Understanding* he wrote:

Knowledge then seems to me to be nothing but the perception of the connexion of and agreement, or disagreement and repugnancy of any of our ideas. In this alone it consists. Where this perception is, there is knowledge, and where it is not, there, though we may fancy, guess, or believe, yet we always come short of knowledge. For when we know that white is not black, what do we else but perceive, that these two ideas do not agree?

Locke's philosophy of perception was responsible for triggering a debate which has failed to produce a satisfactory answer up to the present day. On July 7th 1688 William Molyneux (1656-1698) sent a letter to Locke in which he asked a question regarding what our senses can, or cannot, tell us about the world. In Molyneux's words:

A man, being born blind, and having a globe and a cube, nigh of the same bigness, committed into his hands, and being taught or told, which is called the globe, and which the cube, so as easily to distinguish them by his touch or feeling; then both being taken from him, and laid on a table, let us suppose his sight restored to him; whether he could, by his sight, and before he touch them, know which is the globe and which the cube? Or whether he could know by his sight, before he stretch'd out his hand, whether he could not reach them, tho they were removed 20 or 1000 feet from him?

MATERIALS

Since the earliest civilisations people have been attempting to characterise the objects which surround us in the world. One property defined by modern science is a material's thermal conductivity, which is its ability to conduct heat. Thermal conductivity is defined as the quantity of heat transmitted during a fixed period of time through a given thickness of material due to a fixed temperature difference. Since this property varies from material to material, we find that objects heat or cool at different rates, and sunlight or other heat sources produce different results on different materials. Wood, metal and stone react differently, enabling thermal eyes to see what normal eyes cannot.

right: Metal and plastic demonstrate contrasting thermal properties at this sports facility

opposite above: Metal and stone meet on the face of Big Ben, London's famous clock

opposite below left: Stone, plaster, wood and glass respond differently to sunlight on this timber frame house

opposite below right: Stone, plaster, metal and glass contrast with their wood counterparts on the front face of this house

41

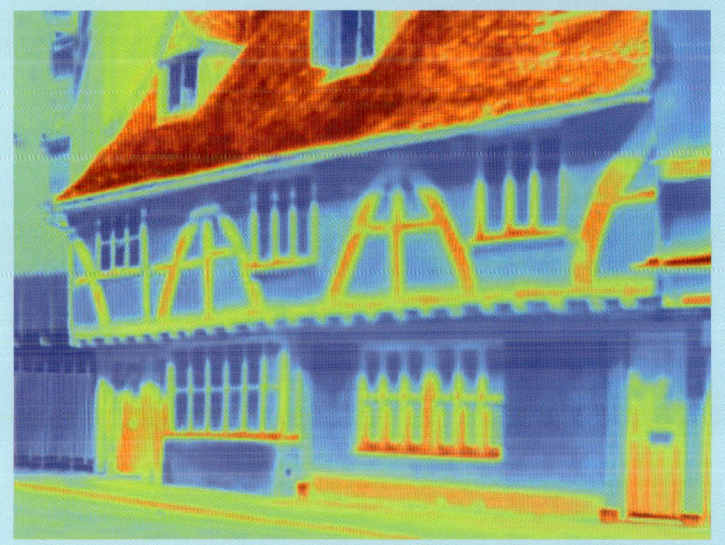

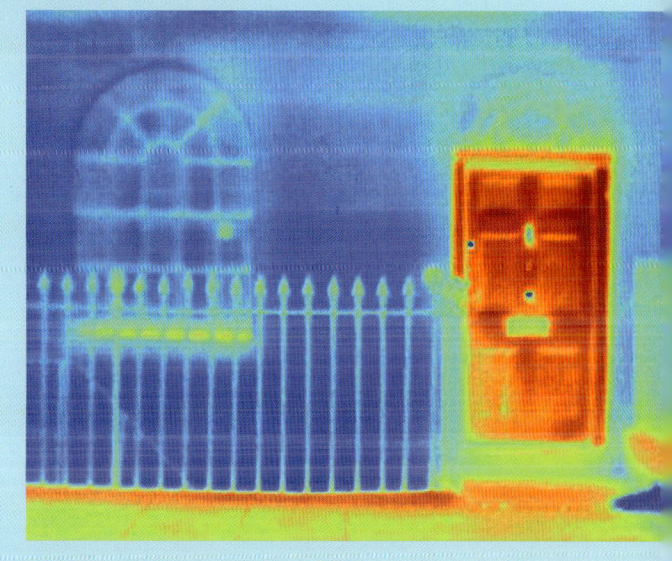

Molyneux's question of whether knowledge acquired by one sense is accessible to another has never been completely answered. Locke and Berkeley, for slightly different reasons, responded negatively, claiming that simultaneous experience from the two senses is necessary in order to make a connection between them. Others have argued the opposite. Unfortunately, after more that 300 years, we are only slightly closer to a definitive answer than Locke was when he replied to his esteemed friend.

George Berkeley (1685 to 1753) profoundly disagreed with the separation of body and mind theory proposed Descartes and Locke, and also with Hobbes' view that only material things and motions exist. In *Of the Principles of Human Knowledge* he wrote:

They who assert that figure, motion, and the rest of the primary or original qualities do exist without the mind, in unthinking substances, do at the same time acknowledge that colours, sounds, heat, cold, and suchlike secondary qualities, do not - which they tell us are sensations existing in the mind alone, that depend on and are occasioned by the different size, texture, and motion of the minute particles of matter. This they take for an undoubted truth, which they can demonstrate beyond all exception. Now, if it be certain that those original qualities are inseparably united with the other sensible qualities, and not, even in thought, capable of being abstracted from them, it plainly follows that they exist only in the mind. But I desire any one to reflect

WATER

Water is a very noticeable element in a thermal world as even the smallest traces, such as evaporation, can have a startling effect. The effects of humidity in a wall or in the soil are as obvious as the presence of fire or light in the visible spectrum. Because of their cooler temperatures, lakes, ponds and rivers stands out against the background. Changes in water temperature produce spectacular effects, turning the water which breaks through the surface of a pond into a raging volcano.

below left: Damp cools a retaining brick wall

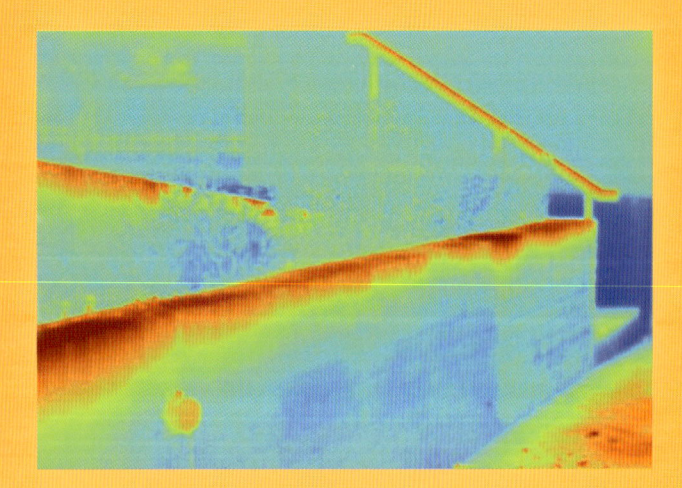

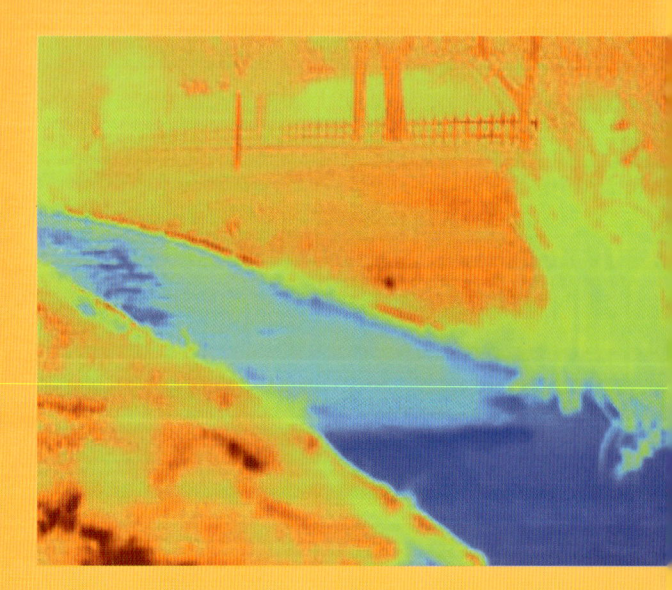

below right: The cool water of a stream contrasts with its warmer surroundings

and try whether he can, by any abstraction of thought, conceive the extension and motion of a body without all other sensible qualities. For my own part, I see evidently that it is not in my power to frame an idea of a body extended and moving, but I must withal give it some colour or other sensible quality which is acknowledged to exist only in the mind. In short, extension, figure, and motion, abstracted from all other qualities, are inconceivable. Where therefore the other sensible qualities are, there must these be also, to wit, in the mind and nowhere else.

For Berkeley, time and space are known to our minds due to God, whose divine provision creates our perception of the world from his. Truth could be found only by stripping away thoughts and language, leaving the forms of pure perception provided to us by God. According to Berkeley

To be is to be perceived.

David Hume (1711-1776) was an Enlightenment philosopher who strongly believed that all knowledge is derived from sensory experience, and that we can have knowledge only of sensation, not of the objects which produce it. He is generally credited with the first systematic theory of how people make

right: Warm water from a pump bursts through the surface of a pond like a small volcano

below right: The cool water of this lake contrasts with the warmth of the swan and the man nearby

below: The water from a fountain cools a group of playful children

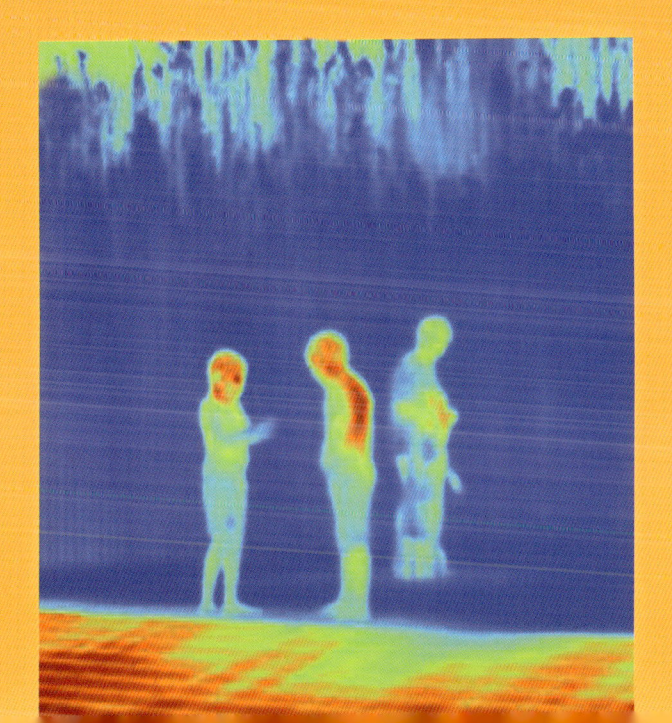

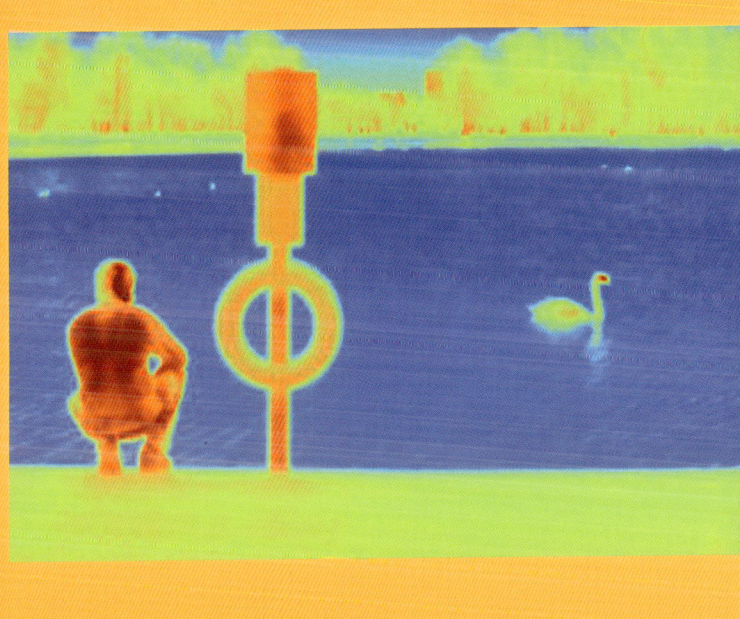

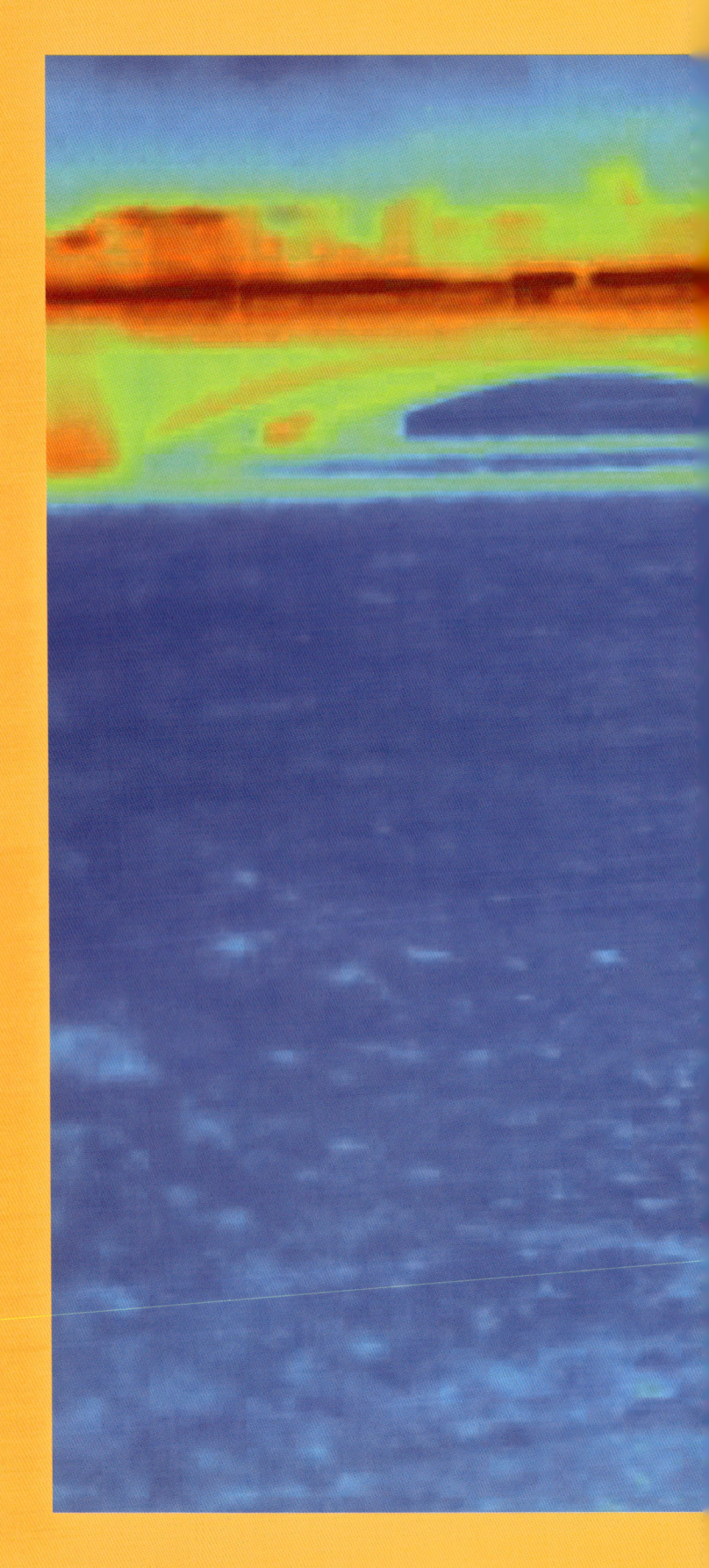

In this view over the River Thames, both natural and man-made objects and the people walking by appear warm in comparison with the cold expanse of water

45

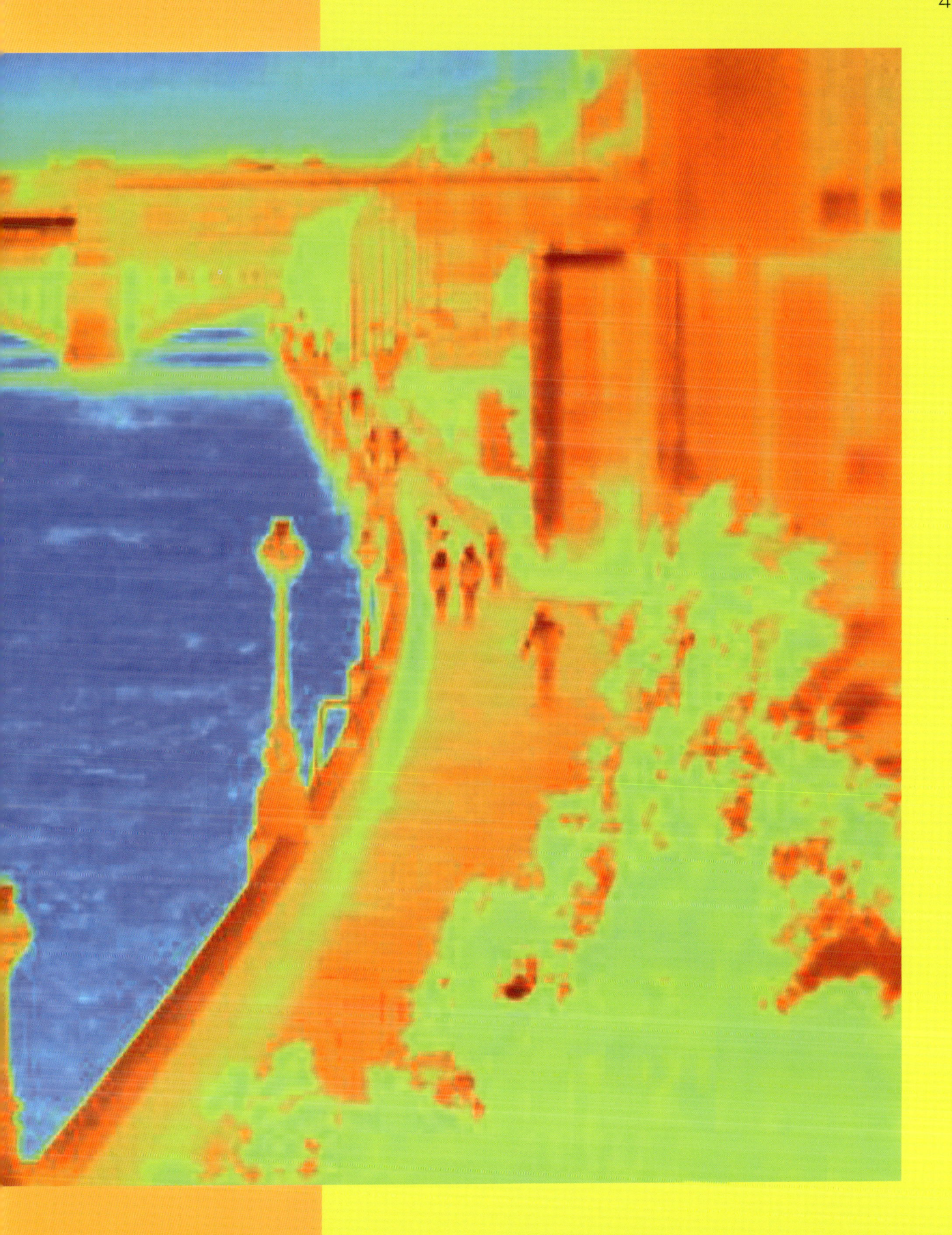

THERMAL EXCHANGE

One curious characteristic of thermal vision is the ability to see the thermal exchange which occurs between objects which have different temperatures. Things imprint themselves on people, and people imprint themselves on things. For a period of time after physical contact, the cooler of the two objects will show signs of the contact with the warmer, and vice versa. Chairs provide a good example as they visually announce the support which they provide to people for many minutes after the person has gone away.

above: A thermal imprint of a person is left behind for several minutes after they have left the chair

below: Thermal vision reveals that someone was sitting in this chair not long ago. The heat imprint of their bare feet is still visible on the floor

opposite above: Two children sit on the sofa, which is gradually warmed by their body heat

opposite below: As the children leave, their thermal imprint remains visible until the sofa returns to the ambient temperature of the room

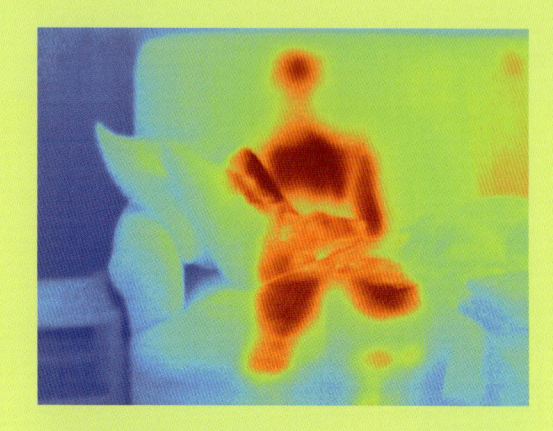

---

inductive inferences, something which in philosophy has come to be known as the problem of induction. According to Hume, people tend to believe that objects behave in a regular manner due to having observed a certain uniformity in nature. People look for causal, thus predictable, patterns and correlations in what they perceive. Hume's strongly empiricist answer to the problem of induction thus places the burden of accurate reasoning squarely on the shoulders of our sensory systems.

Thomas Reid (1710-1796) is another philosopher who contributed to our understanding of perception by proposing a three stage model of its function. In the first stage, objects act upon our bodies by means of causal forces to produce sensations such as colour or temperature. In the second, these sensations produce intuitive thoughts about what may have been the cause of the sensations. In the third, the thoughts, if accompanied by a conviction, become perception. Reid shaped our understanding with the distinction he made between different perceptual stages, which has been widely adopted by modern philosophy and science. Though Reid emphasised the nearly instantaneous and irresistible nature of the transition from sensation to perception, the concept of a series of processing stages has proved immensely influential. The importance of clearly defining the process which bridges the outer physical world and the inner subjective world can perhaps best be understood from Reid's own words in his *Essays on the Intellectual Powers of Man:*

When I smell a rose, there is in this operation both sensation and perception. The agreeable odour I feel, considered by itself, without relation to any external object, is merely a sensation. It affects the mind in a certain way; and this affection of the mind

47

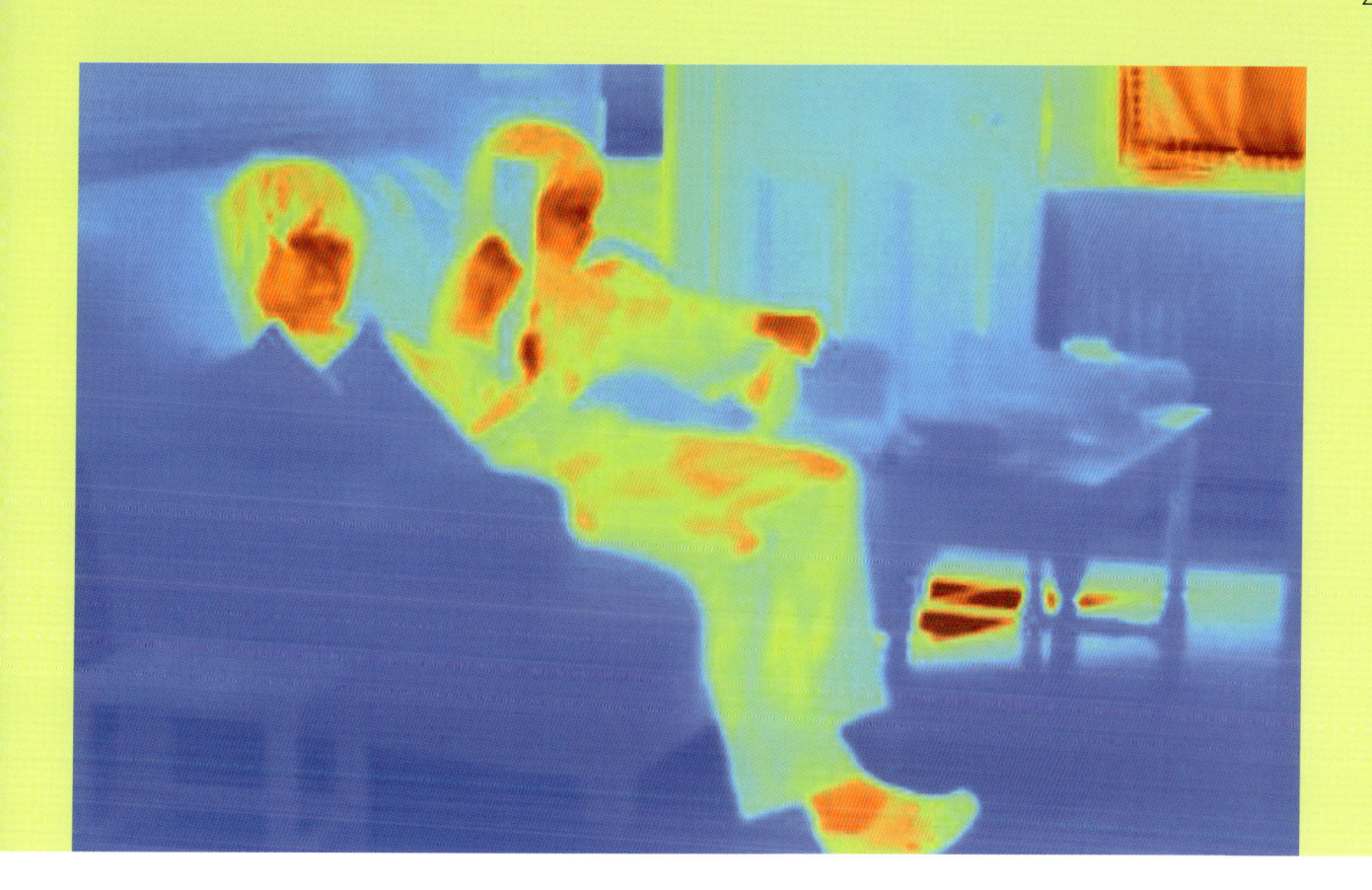

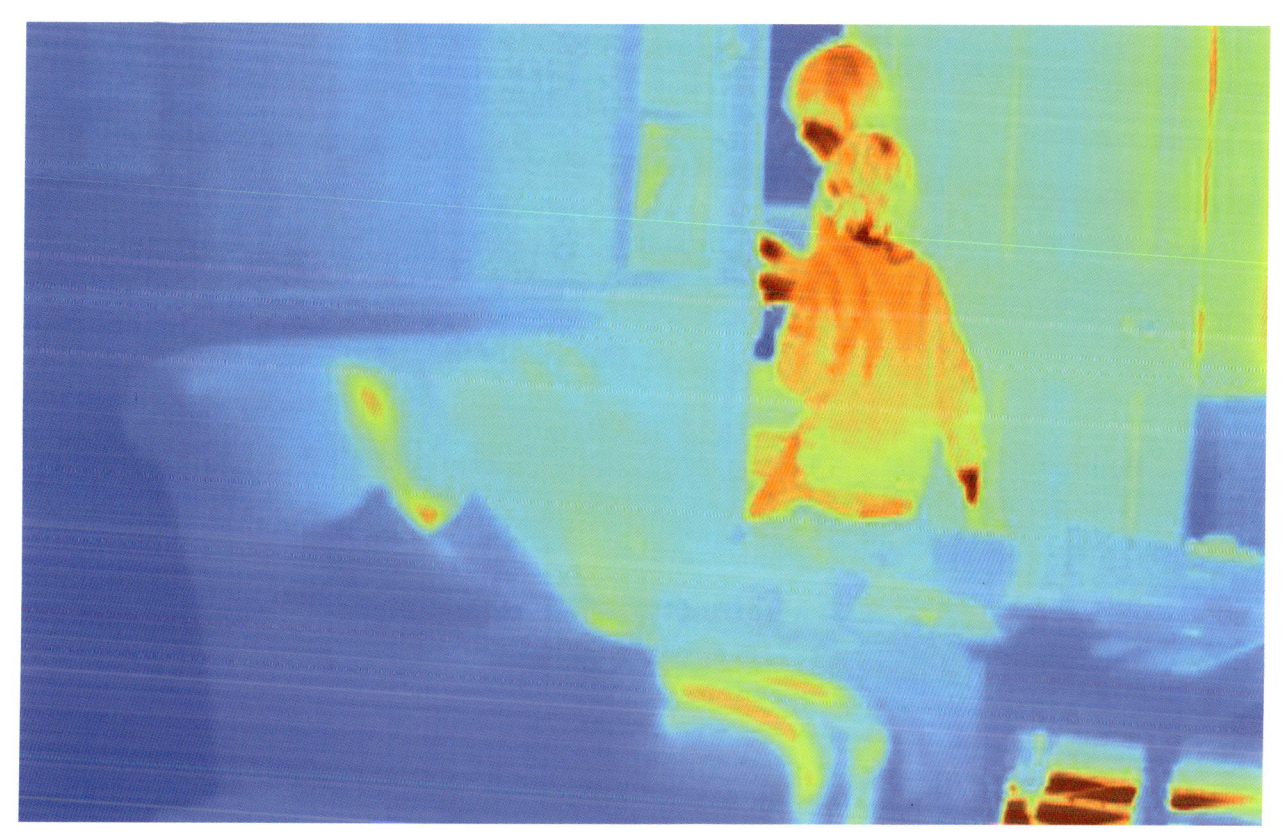

may be conceived, without a thought of the rose, or any other object. This sensation can be nothing else than it is felt to be. Its very essence consists in being felt, and when it is not felt, it is not. There is no difference between the sensation and the feeling of it; they are one and the same thing. It is for this reason, that we before observed, that, in sensation, there is no object distinct from that act of the mind by which it is felt; and this holds true with regard to all sensations.

Let us next attend to the perception which we have in smelling a rose. Perception has always an external object; and the object of my perception, in this case, is that quality in the rose which I discern by the sense of smell. Observing that the agreeable sensation is raised when the rose is near, and ceases when it is removed, I am led, by my nature, to conclude some quality to be in the rose, which is the cause of this sensation. This quality in the rose is the object perceived; and that act of my mind, by which I have the conviction and belief of this quality, is what in this case I call perception.

But it is here to be observed, that the sensation I feel, and the quality in the rose which I perceive, are both called by the same name. The smell of a rose is the name given to both. So that this name hath two meanings; and the distinguishing its different meanings removes all perplexity, and enables us to give clear and distinct answers to questions, about which Philosophers have held much dispute.

VISIBLE VS THERMAL

Appearances change dramatically when switching from the visual spectrum to the thermal spectrum. Uniformity in the thermal world is usually the result of a similarity in materials rather than colours. As a result, differences in material, which are often due to differences in function, are thrust upon us. The location of lighting and electrics is highlighted, the role of metals in providing the basic structure of the modern world is evidenced, and a new world of relationships between things is revealed.

below left: Thermal vision reveals the heat escaping directly through an open window

below right: A metal handrail quickly cools in the breeze

right: Heat reveals the position of the electrical generator on this canal boat

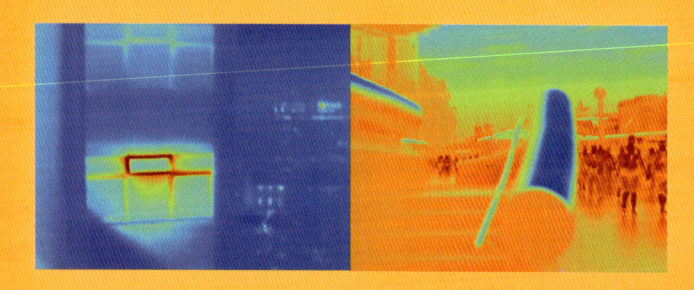

Thus, if it is asked, Whether the smell be in the rose, or in the mind that feels it? The answer is obvious: That there are two different things signified by the smell of a rose; one of which is in the mind, and can be in nothing but in a sentient being; the other is truly and properly in the rose. The sensation which I feel is in my mind. The mind is the sentient being; and as the rose is insentient, there can be no sensation, nor any thing resembling sensation in it. But this sensation in my mind is occasioned by a certain quality in the rose, which is called by the same name with the sensation, not on account of any similitude, but because of their constant concomitancy.

Reid's form of direct realism assumes that the mind is in contact with the external world via the sensory organs. But this begs the question of how accurate the contact is. How far might our sensory organs diverge from the true happenings of the natural world? Reid argued that human perception came close to measuring the true properties of the world, a common sense conviction which he based on the success people have in dealing with the world around them and on innate beliefs about the nature of God's creation.

Perhaps the greatest philosopher of the Enlightenment was Immanuel Kant (1724-1804). His views are key to our modern understanding of perception because he mounted the most influential challenge to direct realism which has ever been attempted. Kant developed a detailed theory based on the idea that

right: The disposed beverage container in the foreground of this picture reveals its contents to prying thermal eyes

below right: Thermal eyes pierce through the darkness of this underground tube tunnel to reveal the workings of the rails and electrical boxes

below: In contrast to our biological vision, thermal eyes see no light emitted from lamps, only the waste heat which is produced in the process

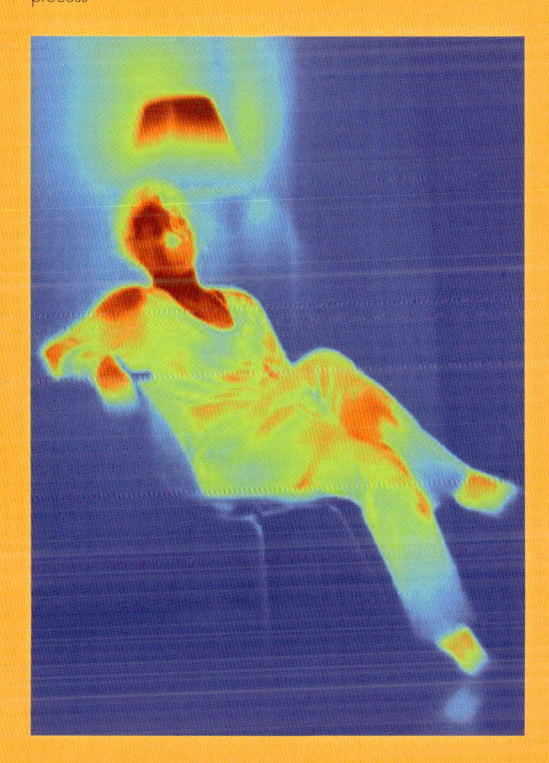

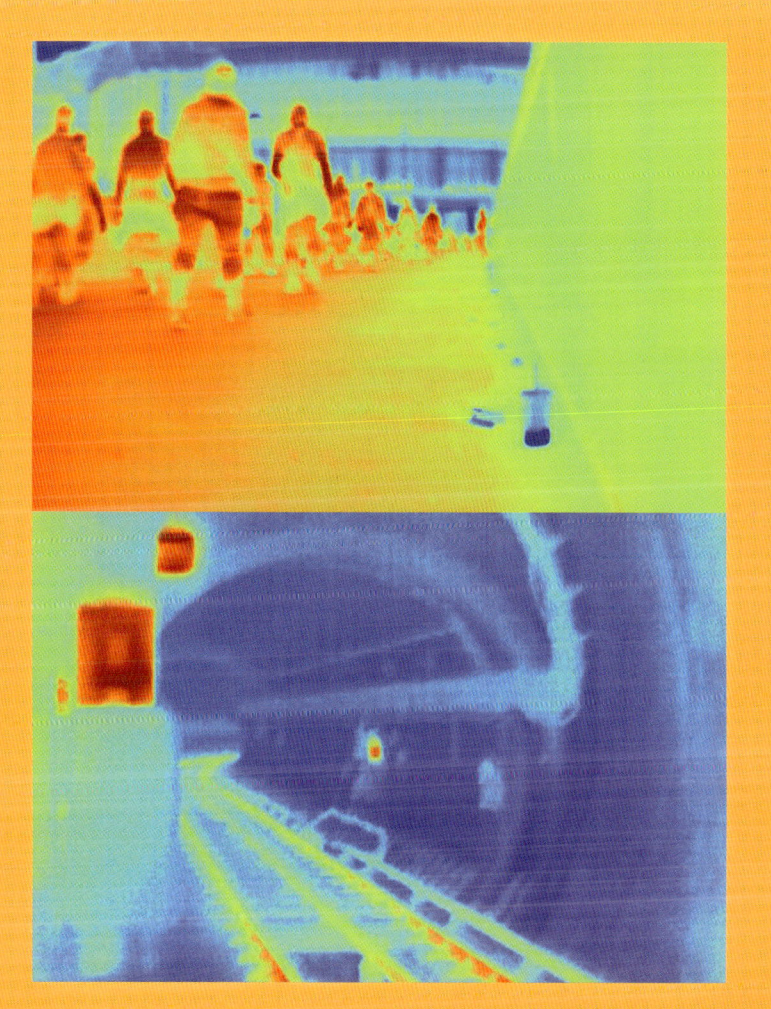

SKY

The sky is perhaps one of the biggest surprises revealed through thermal vision. The blues and greys of our normal visible sky are familiar and engrained within our minds, so it is with a certain shock that we see an intense bright red cloud, filled with moisture. Such unexpected views challenge our understanding of how we perceive the world around us and prompt us to look at our everyday surroundings in a new light.

right: A patchy sky provides a backdrop to the Houses of Parliament in London

opposite: Hot, humid clouds appear grey or white to our biological vision but become red when viewed with thermal eyes, offering an insight into the nature of the cloud formation

time and space are not independent concepts of the world around us, but rather forms of sensibility that are a priori necessary conditions within us. He proposed a fundamental new interpretation of sensory experience based on the idea that the objective order of nature and the causal necessity that operates within it are structures of the mind. Kant speaks of the "thing in itself" or the "transcendental object" in order to separate the reality of objects from what we perceive of them, something which is difficult to achieve in practice due to the overwhelming order imposed by our mind as it attempts to create structure and abstraction from raw sensory information. Kant described his theory as a "Copernican Revolution" which inverted our model of the world, replacing the concept of external objects which structure sensory representations with the concept of sensory representations which structure external objects.

A cornerstone of Kant's theory was the idea that sensation and perception are temporal in nature, meaning that the quantity that we call time acts as a marker which we use to order and interpret the sensations provided to our minds by our sensory organs. According to Kant, our minds could not order sensory images if it were not possible to label each one as separate and distinct. Furthermore, without the innate ability to label images that occur at different moments in time it would not be possible to establish the persistence of objects over time, to understand the fact that images of the same object can change with changes of viewing angle, or the fact that properties of the object can change with the passing of time. Kant argued that a temporally extended series of experiences is the very basis of self-consciousness, which he referred to as the "necessary unity of consciousness", or what we might today call "self".

Another important principle of Kant's theory was the idea that space, like time, is an innate and a priori structure of our mind which is used to make sense of the information provided by our sensory organs. Kant maintained that we identify the objects around us due to correlations and causations which we note in both time and space. We label something as being a single worldly object due to the nearness in either time or space of the sensory impressions caused by the object. For example, visual images would be credited with having been caused by the same table if the images were either close in time or spatially similar even after long periods of time. In *Critique of Pure Reason* of 1781 Kant wrote:

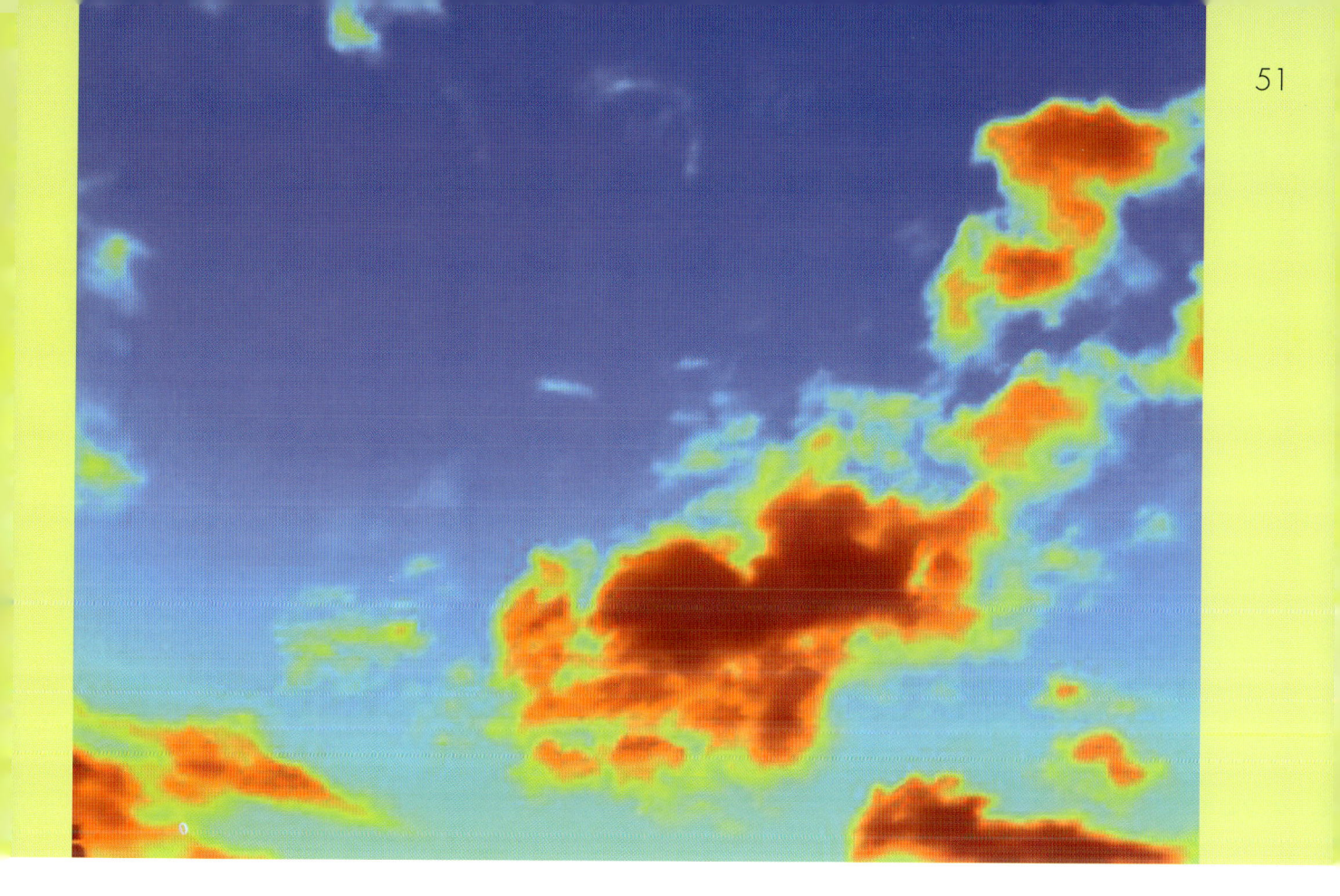

If I take away from the representation of a body that which the understanding thinks in regard to it, substance, force, divisibility, etc., and likewise what belongs to sensation, impenetrability, hardness, colour, etc., something still remains over from this empirical intuition, namely extension and figure. These belong to pure intuition, which, even without any actual object of the senses or sensation, exists in the mind a priori as a mere form of sensibility.

Perception in the Modern World

In more recent times philosophers have expanded our understanding of sensation and perception by considering its content and structure in ever greater detail. A major 20th century movement which shaped our understanding was that of phenomenology.

Franz Brentano (1838-1917) is generally considered to be the father of a form of descriptive psychology which eventually came to be known as phenomenology. It analyses the structure of consciousness from a first-person intentional perspective, focusing on the detailed content of perception,

Thermal eyes can see the sun even through thick storm clouds

53

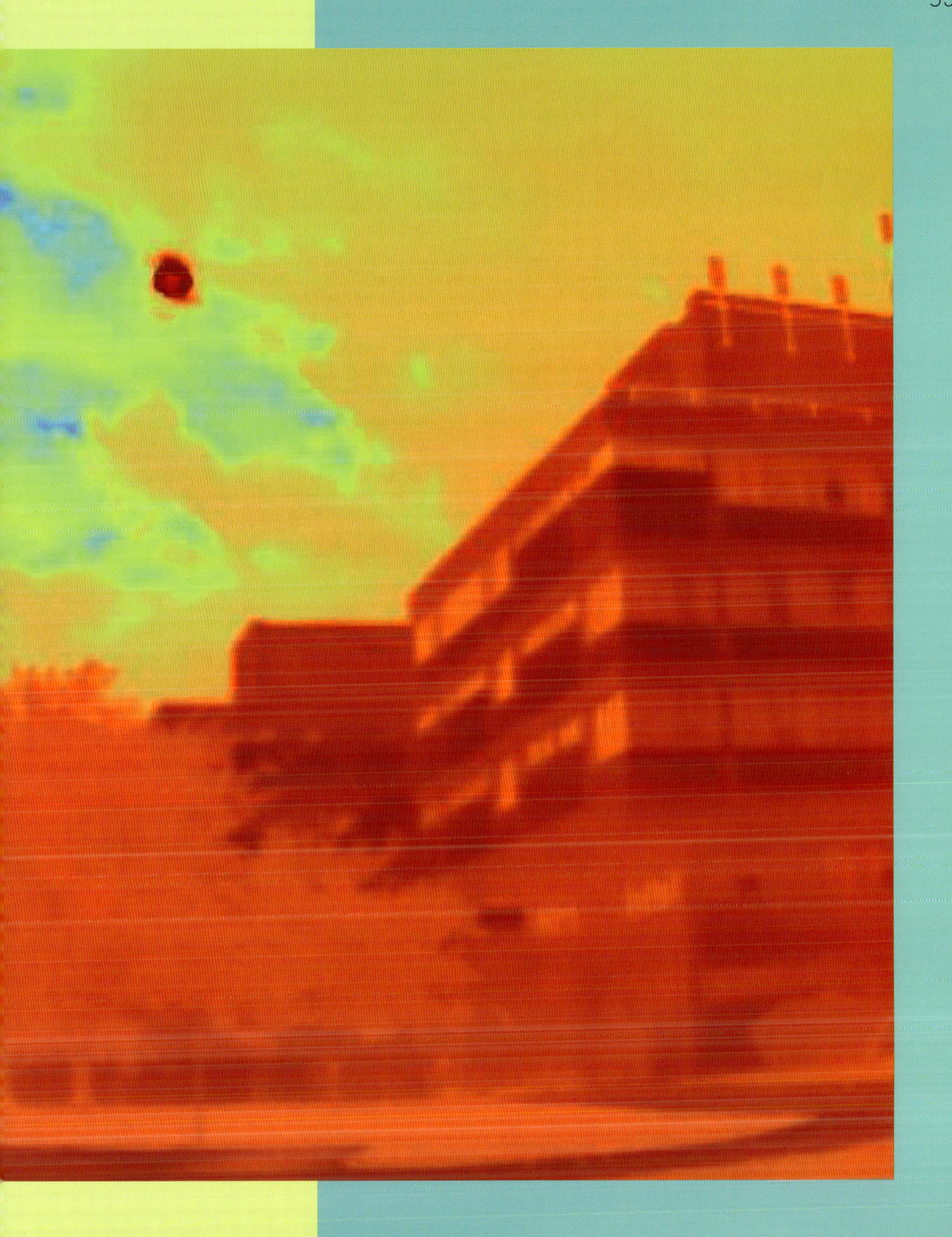

thought, memory, emotion, language and social activity. It considers the relationships which we immediately and spontaneously assume between the contents of perception and ourselves. In 1874 Brentano wrote in *Psychology from an Empirical Standpoint* that

Every mental phenomenon is characterized by what the scholastics of the middle ages called the intentional (or mental) inexistence of an object, and what we might call, through not wholly unambiguously, reference to a content, direction toward an object (which is not be understood here as meaning a thing), or immanent objectivity. Every mental phenomenon includes something as object within itself, although they do not all do so in the same way. In presentation something is presented, in judgement something is affirmed or denied, in love loved, in hate hated, in desire desired and so on.

The phenomenological approach considers perceptual experience to be the source of all knowledge, thus it focuses on details such as the size of an object, its orientation and colour, and how these details change over time. The goal is to achieve awareness about the pre-reflective contents of perception and the effect that our own body and actions have on these. According to Brentano, our directed associations define the sensation as much as any other property. In *Psychology from an Empirical Standpoint* he writes:

SALISBURY CATHEDRAL

Thermal vision provides unique insights into the world of religious architecture. Using thermal eyes, Salisbury Cathedral appears warm and inviting from outside but proves a much colder place inside. Within the thick walls of the cathedral, the cool temperatures appear as a sea of blue against which worshippers reveal their presence by the yellow and orange of their warmth. As in the visible spectrum, thermal energy from outside enters through the windows and produces splendid effects within the cathedral walls. Unlike the visible spectra, however, the energy is relatively untouched by the colours of the stained glass windows.

right: Inside the thick stone walls, the only sources of heat are people and the security cameras above

opposite above left and right: The stained glass windows warm up as the sun shines through

opposite below: The decorative stone parts are more conductive of heat than the bricks of the wall and so appear warmer

In cases such as this we always have a presentation of a definite spatial location which we usually characterise in relation to some visible and touchable part of our body. We say that our foot or hand hurts, that this or that part of the body is in pain. Those who consider such a spatial presentation something originally given by the neural stimulation itself cannot deny that a presentation is the basis of the feeling. But others cannot avoid this assumption either. For there is in us not only the idea of a definite spatial location but also that of a particular sensory quality analogous to colour, sound and other so-called sensory qualities, which is a physical phenomenon and which must be clearly distinguished from the accompanying feeling. If we hear a pleasing and mild sound or a shrill one, harmonious chord or a dissonance, it would not occur to anyone to identify the sound with the accompanying feeling of pleasure or pain. But then in cases where a feeling of pain or pleasure is aroused in us by a cut, burn or a tickle, we must distinguish in the same way between a physical phenomenon, which appears as the object of external perception, and the mental phenomenon of feeling, which accompanies its appearance, even though in this case the superficial observer is rather inclined to confuse them.

The search for intentional and pre-reflective content launched by the phenomenologists Franz Brentano and Edmund Husserl (1859-1938) also served as inspiration for a group of psychologists who formulated influential theories about sensation and perception. The Gestalt psychologists developed

above: The different materials (wood, metal and stone) that make up the exterior of the cathedral all respond differently to heat

below: At the back of the room the warm wooden door is clearly visible as it contrasts with the cooler stone

experimental methods based on the principle that the brain is holistic, parallel and self-organising. Gestalt, which in German can be understood to mean "shape" or "form", refers to our innate and a priori ability to organise perceptions as unitary wholes. Gestalt psychology focused on the human ability to perceive complete time sequences, shapes and figures, rather than individual details such as angles, lines or colours. Their experiments investigated how people's perceptions consisted of orderly simplifications of the complex perceptual fields which occur in everyday life. The guiding principle was that of "prägnanz" meaning "conciseness". Gestalt psychologists defined "prägnanz" in terms of simple laws such as closure, similarity, proximity, symmetry, continuity and common fate, which each represent an important simplification that seems to occur almost automatically and instantaneously within our sensory systems and mind.

The 20th century phenomenologist and existentialist Maurice Merleau-Ponty (1908-1961) moved the focus of phenomenological investigation even further in the direction of the human body. While working with individuals who suffered from various physical disabilities, he became aware not just of the intentionality of consciousness, but also of its fundamental corporeality. Merleau-Ponty became convinced that the human body acted as the omnipresent condition of all sensory experience. More so than with

LONDON

Like any major city, and perhaps more so than most, London displays a kaleidoscope of forms and materials which have emerged from the layers of its architectural heritage. Thermal vision quickly reveals differences in the use of materials among the structures belonging to the various historical periods. Thermal eyes also reveal some of the tricks of the architect's trade, such as the use of different materials for visibly similar structures, as in the case of St. Paul's dome or the front of a Soho shop. If studied carefully, one can even tell that the well-known statue of Eros is made of aluminium, rather than bronze or steel.

right: The metal statue of Eros warms in the afternoon sunlight, vying for attention with an imposing sky of warm clouds

opposite above left: Tower Bridge stands out against the cool colours of the sky and river

opposite below left: Thermal perception reveals the point at which wood and lead take over from stone on the cupola of St. Paul's Cathedral

most other philosophers, Merleau-Ponty's approach contrasts strikingly with the body-mind duality of René Descartes or the causal and coherence based systems of the Enlightenment philosophers. Typical of Merleau-Ponty's world view is a passage from his best known work, *Phenomenology of Perception*:

The "sensible quality", the spatial limits set to the percept, and even the presence or absence of a perception, are not de facto effects of the situation outside the organism, but represent the way in which it meets stimulation and is related to it. An excitation is not perceived when it strikes a sensory organ which is not "attuned" to it The function of the organism in receiving stimuli is, so to speak, to "conceive" a certain form of excitation. The "psychophysical event" is therefore no longer of the type of "worldly" causality, the brain becomes the seat of a process of "patterning" which intervenes even before the cortical stage, and which, from the moment the nervous system comes into play, confuses the relations of stimulus to organism. The excitation is seized and reorganised by transversal functions which make it resemble the perception which is about to arouse. I cannot envisage this form which is traced out in the nervous system, this exhibiting of a structure, as a set of processes in the third person, as the transmission of movement or as the determination of one variable by another. I cannot acquire detached knowledge of it. In so far as I guess what it may be, it is by abandoning the body as an object, partes extra partes, and by going back to the body which I experience at this moment, in the manner, for example, in which my hand moves round the object it touches,

right: A mix of incandescent and neon lights attract customers to this café bar

far right: The angle of the sun warms one side of the awnings of a department store

below right: The glass and steel structure of London's City Hall

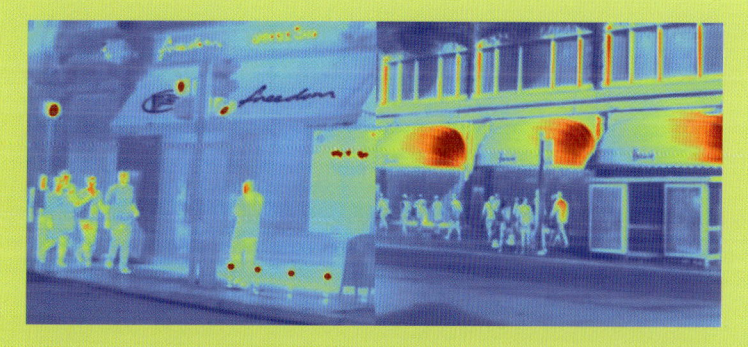

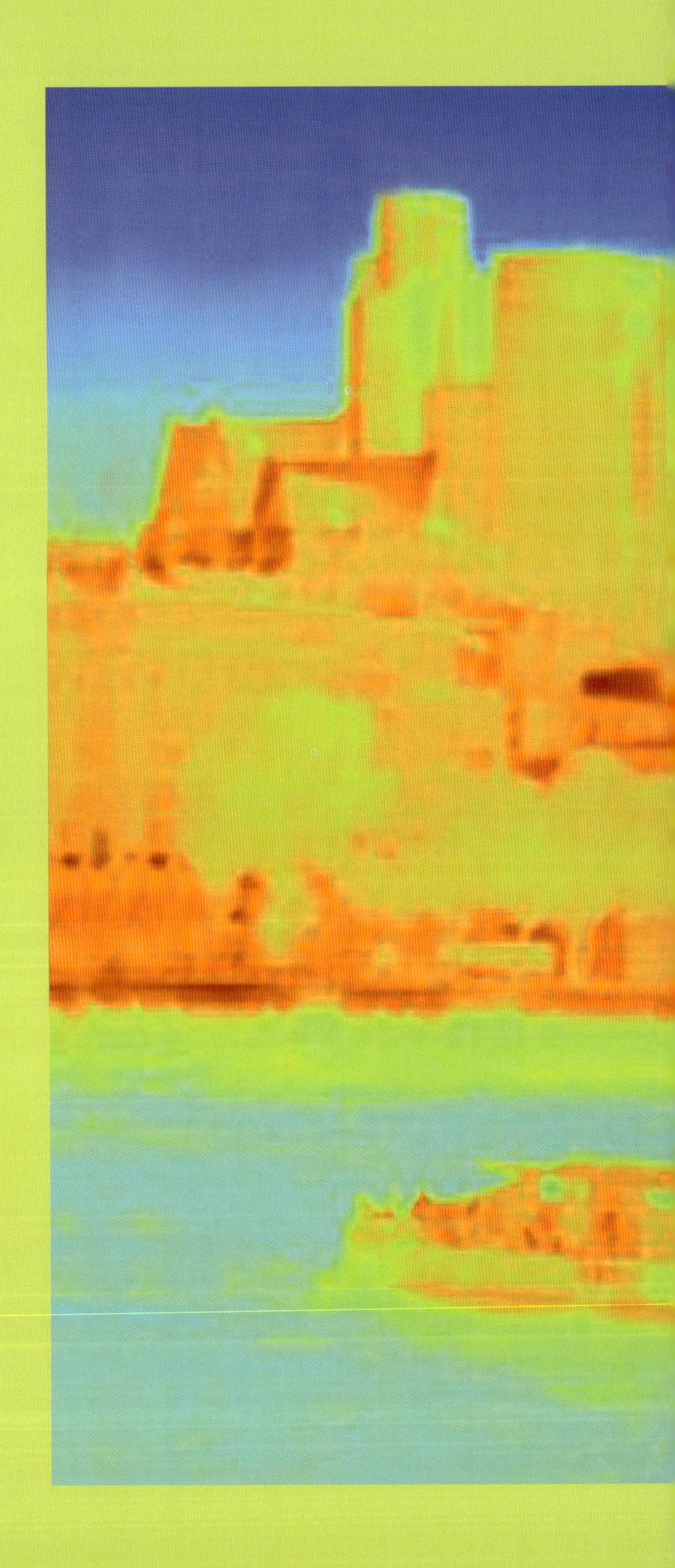

London city skyline with the River Thames in the foreground

59

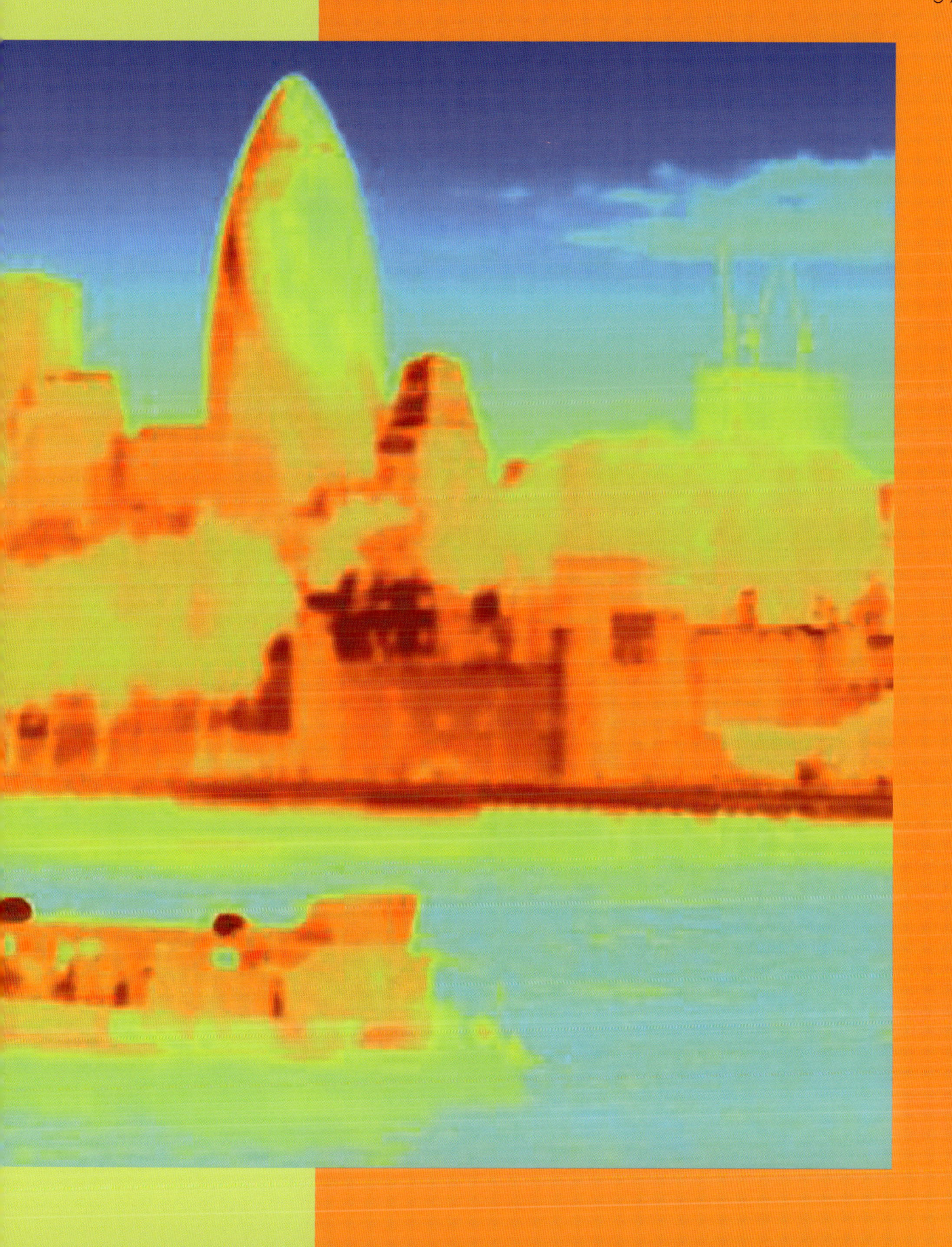

MILLENNIUM BRIDGE

The Millennium Bridge is a fascinating structure when viewed through thermal eyes. On a warm summer's afternoon its metal and stone construction are beautifully highlighted and its location on the River Thames provides panoramic views over the heart of London. The thicknesses of the metal and its contact with both the river and with people all conspire to produce surprising contrasts when viewed through thermal eyes.

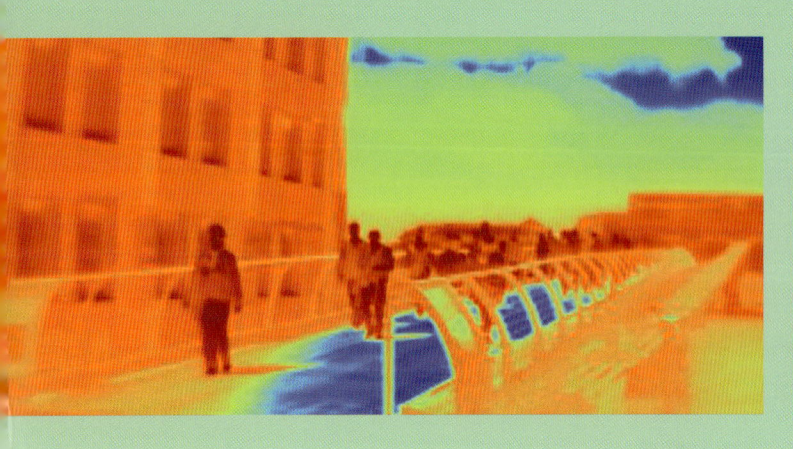

anticipating the stimuli and itself tracing out the form which I am about to perceive. I cannot understand the function of the living body except by enacting it myself, and except in so far as I am a body which rises towards the world.

Merleau-Ponty's phenomenology was rich in detailed accounts of individual instants of perception and of individual encounters with the physical world. His was a search for meaning within the texture of the sensory and the immediate, rather than the pondered and rational, a point of view which he emphasised in *Phenomenology of Perception*:

Once more, reflection – even the second-order reflection of science – obscures what we thought was clear. We believed that we knew what feeling, seeing and hearing were, and now these words raise problems. We are invited to go back to the experiences to which they refer in order to redefine them.

In contrast to what most philosophers have written about perception, psychologist James Jerome Gibson (1904-1979) developed a theory of visual perception that considered it to be simply the name we give to the direct detection of environmental invariances. Gibson's theory states that humans perceive the environment directly, no more, and no less. Unlike models of perception in which the mind is believed to selectively manipulate and interpret sensory information, Gibson's theory states instead that the useful invariant structures are already present in the stimuli and that our biological receptors are only devices which have evolved to respond to those structures. In *The Senses Considered as Perceptual Systems* he wrote:

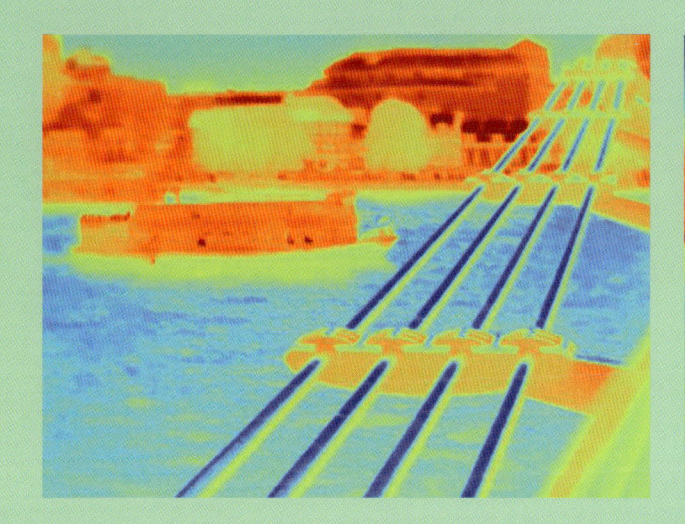

opposite left and right: The thin layer of aluminium that forms the walking platform of the bridge is cooled by the breeze

below: An view of the bridge looking towards St. Paul's Cathedral

above left: The steel cables of the bridge's suspension system are cooled by the wind and so appear very dark blue against the river below

above right: The concrete base of the bridge remains warmer than the metal superstructure above

If this formula is correct, the input of the sensory nerves is not the basis of perception as we have been taught for centuries, but only the basis for passive sense impressions. These are not the data of perception, not the raw material out of which perception is fashioned by the brain. The active senses cannot be simply the initiators of signals in nerve fibers or messages to the brain; instead they are analogous to tentacles and feelers. And the function of the brain when looped with its perceptual organs is not to decode signals, nor to interpret messages, nor to accept images. These old analogies no longer apply. The function of the brain is not even to organize the sensory input or to process the data, in modern terminology. The perceptual systems, including the nerve centres at various levels up to the brain, are ways of seeking and extracting information about the environment from the flowing array of ambient energy.

During the 20th century, science shed much light on the structure and function of sensory systems, and previously hidden secrets were revealed, from the operational principles of individual receptors to the parallel and distributed networks of the brain. A model of a neural network is now a simple thing to construct on a home computer, and anyone who regularly watches the evening news will be better informed about the workings of human perception than the philosophers and scientists of the past.

One 20th century philosopher who noted the deep permeation of scientific logic into modern systems of philosophical enquiry was Peter Frederick Strawson (1919-2006). He suggested that the way we now approach the topic of perception and the questions which we ask lead directly to a causal dependence on the objects perceived. In contrast to Gibson's belief in scientific structure in the external world, Strawson argued that such structures were simply the result of the scientific logic which we currently adopt. Strawson's descriptive metaphysics can be considered both a warning about the dangers of overly scientific enquiry and a reminder of the many aspects of the body-mind problem which still remain to be solved before the world of the external and internal can be reconciled. In *Analysis and Metaphysics: An Introduction to Philosophy* he wrote:

The notion of the causal dependence of the experience enjoyed in sense-perception on features of the objective spatio-temporal world is implicit from the very start in the notion of sense-perception, given that the latter is thought of as generally issuing in true judgements about the world. It is not something we discover with the advance of science, or even by refined philosophical argument. It is conceptually inherent in a gross and obvious way in the very notion of sense perception as yielding true judgements about an objective spatio-temporal world. Hence any philosophical theory which seeks to be faithful to our general framework of ideas, our general system of thought, must provide for this general notion of causal dependence. It must, to this extent at least, be a causal theory of perception.

Some worrisome conclusions which result from the application of modern scientific method to the problem of sensory perception were noted by Friedrich August von Hayek (1899-1992). Hayek developed a model of perception based on state-of-the-art scientific knowledge about the workings of the human nervous system, then investigated some of the logical conclusions which could be drawn based on the model. According to Hayek, perception was the outcome of classification stages performed by neurons as nervous impulses progressed from the sensory periphery to the cortex. With the clustering and classification of stimuli established as the basis for sensory experience, Hayek wrote in

63

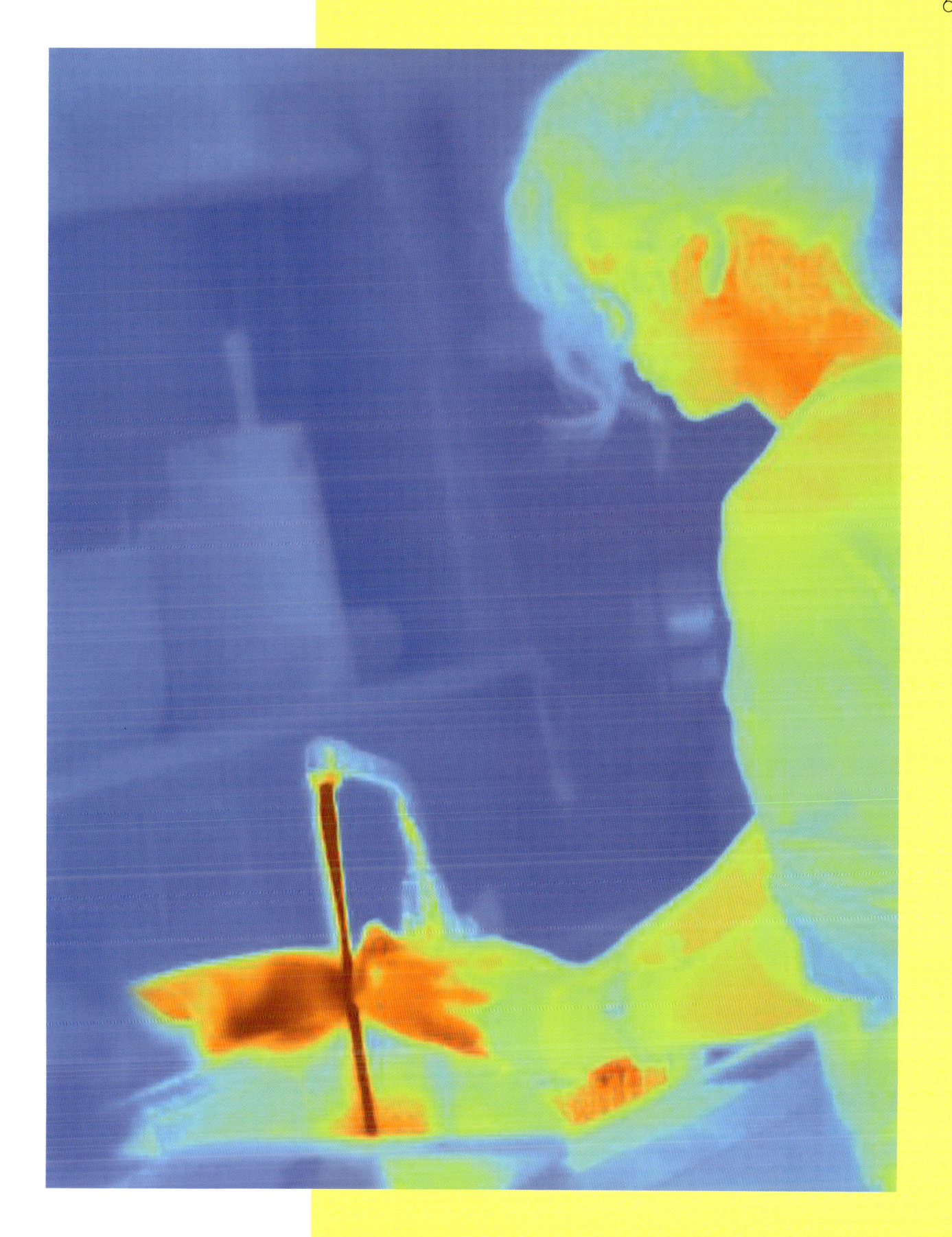

The Sensory Order: An Inquiry into the Foundations of Theoretical Psychology that

If sensory perception must be regarded as an act of classification, what we perceive can never be unique properties of individual objects but always only properties which the objects have in common with other objects. Perception is thus always an interpretation, the placing of something into one or several classes of objects. An event of an entirely new kind which has never occurred before, and which sets up impulses which arrive in the brain for the first time, could not be perceived at all.

All we can perceive of external events are therefore only such properties of these events as they posses as members of classes which have been formed by past "linkages". The qualities which we attribute to the experienced objects are strictly speaking not properties of that object at all, but a set of relations by which our nervous system classifies them or, to put it differently, all we know about the world is of the nature of theories and all "experience" can do is to change these theories.

This means also that what we perceive of the external world are never either all the properties which particular objects can be said to possess objectively, nor even only some of the properties which these objects in fact do posses physically, but always only certain "aspects", relations to other kinds of objects which we assign to all elements of the classes in which we place the perceived objects. This may often comprise relations which objectively do not at all belong to the particular object, but which we merely ascribe to it as the member of the class in which we place it as a result of some accidental collocation of circumstances in the past.

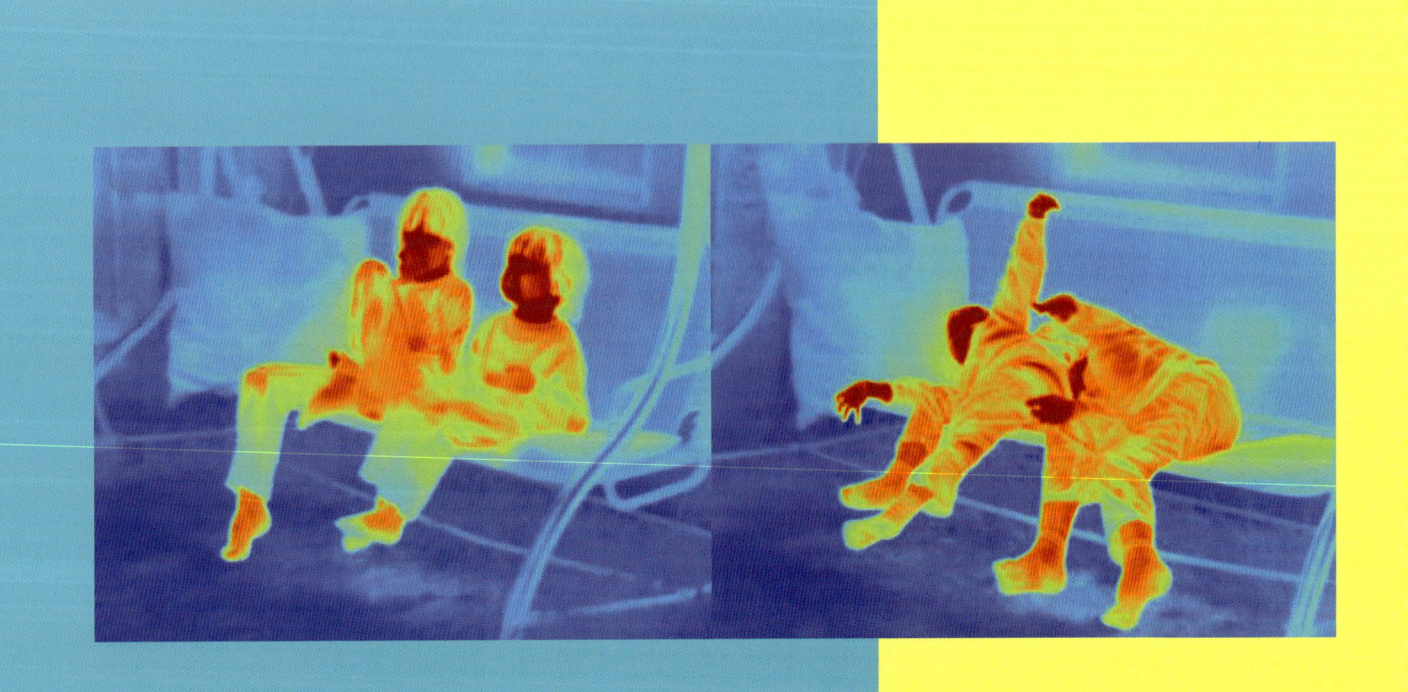

Beyond exploring the limitations of modern science, 21st century endeavour has also seen a mixing of the forms of enquiry which originated within the philosophical and scientific disciplines, leading to more multidisciplinary and comprehensive approaches. Typical of this new way of thinking is the work of philosophers such as Shaun Gallagher, who has drawn from phenomenology and cognitive science to carefully articulate the concept of embodied cognition. In *How the Body Shapes the Mind*, he writes:

In the beginning, that is, at the time of our birth, our human capacities for perception and behaviour have already been shaped by our movement. Prenatal bodily movement has already been organized along the lines of our own human shape, in proprioceptive and cross-modal registrations, in ways that provide a capacity for experiencing a basic distinction between our own embodied existence and everything else. As a result, when we first open our eyes, not only can we see, but also our vision, imperfect as it is, is already attuned to those shapes that resemble our own shape. More precisely and quite literally, we can see our own possibilities in the faces of others. The infant, minutes after birth, is capable of imitating the gesture that it sees on the face of another person. It is thus capable of a certain kind of movement that foreshadows intentional action, and that propels it into a human world.

Gallagher's embodied cognition adopts aspects of phenomenology, psychology, psychophysics and neuroscience to define how our physical human body influences the workings of our perception and thought. Key aspects such as our shape, strength and movement abilities all serve to direct our consciousness and to ground and scale the sensory streams which we encounter in everyday life. Without our own bodies, to what would we compare the size of an object? How fast would we consider a speed of movement to be? It is the living and breathing reference system which we all possess that grounds and scales the world around us.

Before closing this short review of the multiple interpretations of perception, it is worth mentioning the work of Thomas Nagel. Nagel is an advocate of the idea that subjective experience cannot, at least with current scientific understanding, be reduced simply to brain activity. He has considered in

page 63: Hot water from the tap runs over the dishes and the woman's hands, transferring heat

opposite left and right: Two children play outdoors on a garden swing. As they sit together, heat transfers to the padded seat of the swing

right: Two children stand out against the cool colours of their surroundings

page 66: In this living room scene the stereo on the left can be seen emitting heat energy

page 67: Three children explore their new thermal world

detail the difficulties involved in attempting to transport the subjective experiences of one person to another, or those of one creature to another. Inventive examples such as his 1974 article *What is it Like to Be a Bat?* have highlighted the deeply intractable nature of subjective experience, and thus the difficulty of producing a scientific explanation of the subjective experiences of living creatures. Nagel's great merit has been to remind us that no amount of philosophical or scientific enquiry can put us in the proverbial shoes of another.

As he has argued in works such as *The View From Nowhere*, it is perhaps the ultimate destiny of all philosophical enquiry to conclude that given the impossibility of being another living creature, perhaps the best we can do is develop the abstract and distanced understanding that all forms of science seek. According to Nagel, an integral part of our thought process is the ability to think about the world in terms which transcend our own experience, considering the world from a vantage point that is, in Nagel's words, "nowhere in particular". As a race of intelligent creatures we appear destined to continue our quest for an explanation of perception, though Nagel leaves us with a word of warning in *The View From Nowhere*:

In pursuing objectivity we alter our relation to the world, increasing the correctness of certain of our representations of it by compensating for the peculiarities of our own point of view. But the world is in a strong sense independent of our possible representations, and may well extend beyond them.

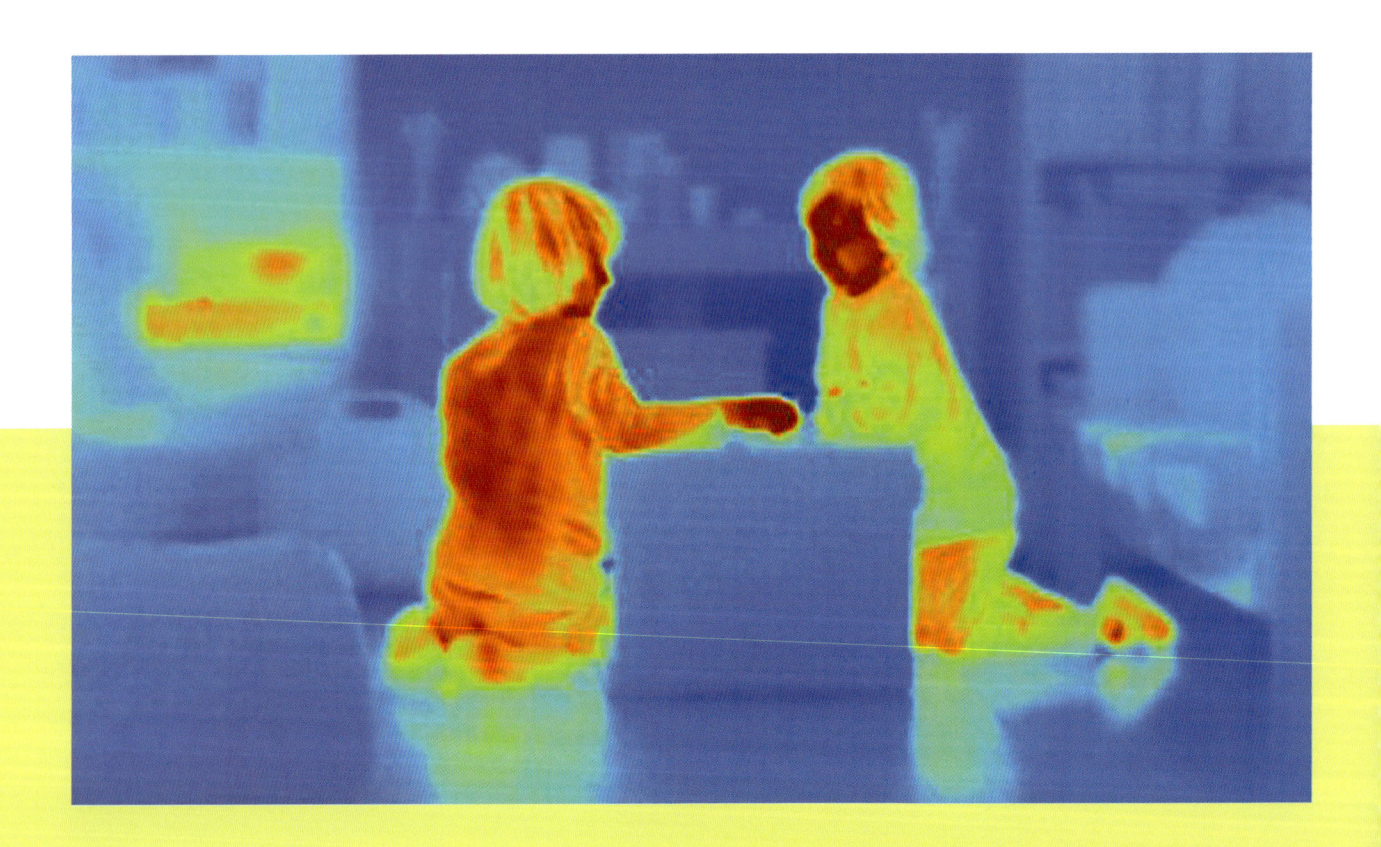

67

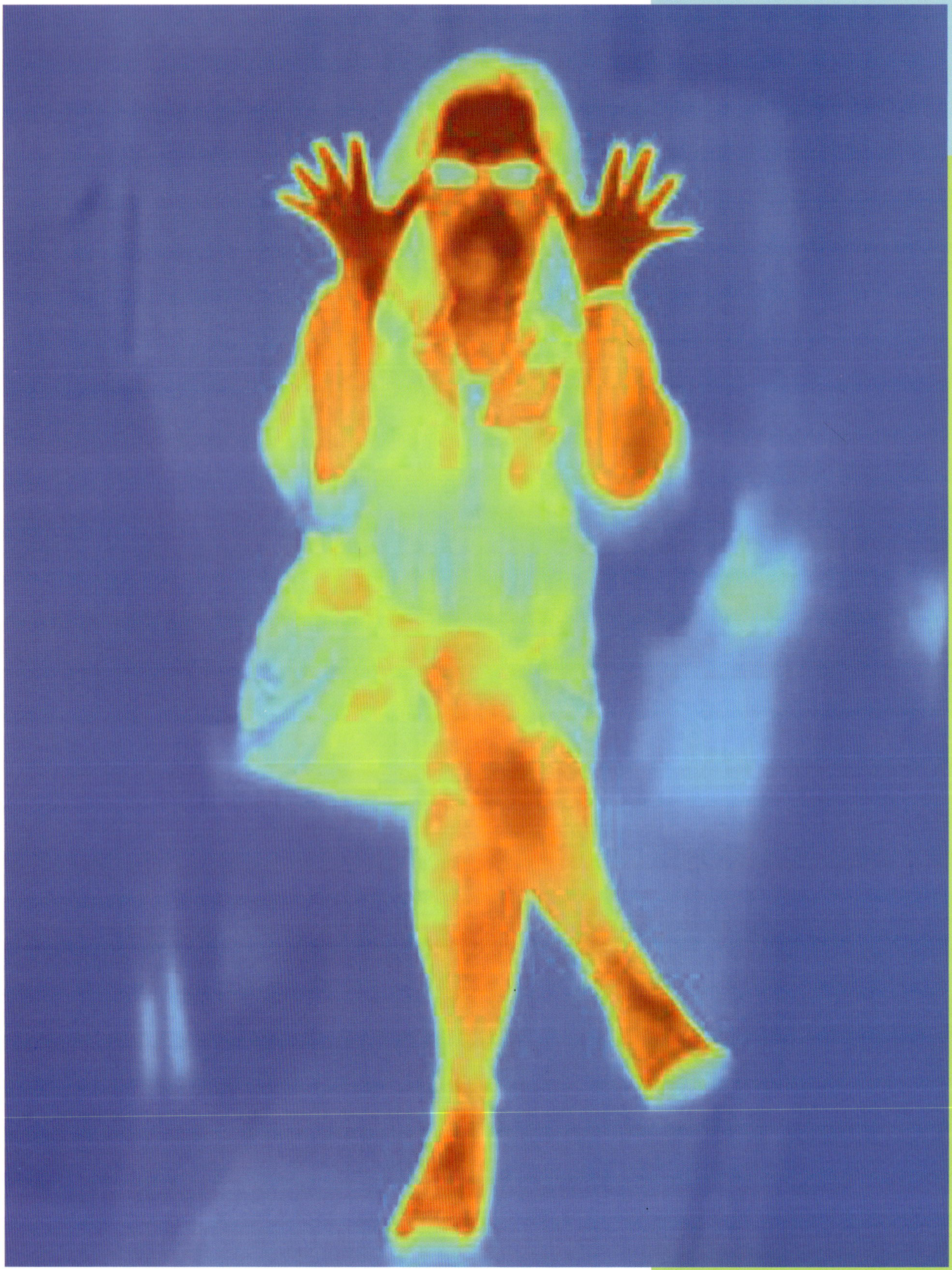

PSYCHEDELICS

It can be said that perception is a subject which has never quite been fully explained by philosophical and scientific endeavour. The gap between the objective world and the subjective world may have closed slightly over the generations, but, like a philosophical Grand Canyon, it is still there. Rational thought has not offered us the elusive panacea of an ultimate and definitive knowledge of how we are to understand and live with the "things in themselves". Contact with the external world leaves us uncertain, unsatisfied, possibly even insecure. Have we really correctly perceived the things which are in the world? When we do things, is our knowledge based on fact or smoke and mirrors? When we peek through the hole in the fence, do we see everything which is on the other side? For those who wish to explore the world and design new and exciting things, what can provide assistance other than rational philosophical or scientific thought? Where should insights be sought?

Since ancient times, the desire to transcend the ordinary and explore the nature of the "things in themselves" has lead to efforts to modify the sensitivity and function of our biological perceptual systems. Sensory deprivation, sensory overload, bodily harm, meditation, training, routine and diet have all been attempted, often in conjunction with ritualistic or religious ceremony. For many, the temptation to enhance our natural biological systems so as to see the unseeable and perceive the unperceivable is too great to resist. One particularly powerful approach which seems to have been practiced by all cultures throughout all of history has been the ingestion of chemical substances which alter perception. Such chemical shortcuts have transported and transformed many individuals, including many of the artistic, religious, philosophical or scientific visionaries whose names can be found filling the pages of our history books.

One substance which has been used to enhance perception since the beginning of record keeping is

UNDERGROUND STATION

Underground stations provide many thermal attractions, such as the track, signalling equipment, advertising and architecture. Warm rooms, roofs, electrical boards and people all conspire to make an underground station a strange and unfamiliar thermal world for travellers.

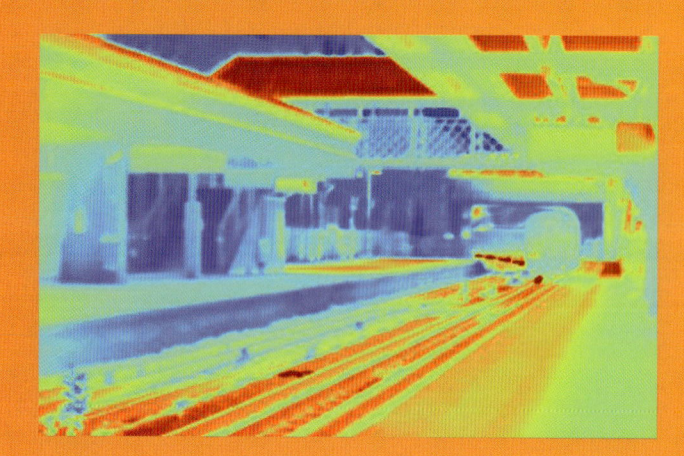

page 68: The different materials of this woman's glasses and watch stand out clearly

page 69: A woman walks by in the rain with her umbrella and shopping bags. The rain cools her surroundings and the statue behind, but her body temperature remains warm

hemp, or *Cannabis sativa*. Described in detail in Chinese texts from about 2737 BC and cultivated from at least 8000 BC, it is believed to be one of humanity's first agricultural crops. The earliest references to the perception altering effects of cannabis can be found in one of the earliest religious documents, the *Atharva Veda* (2000-1400 BC). One of the four sacred *Vedas* of Hinduism, the *Atharva Veda* refers to cannabis as *bhanga*, a sacred grass used in ritual to produce ecstatic states. Practical experimentation to alter perception by means of cannabis would appear to predate all known Western philosophical enquiry.

The Hindu *Rig Veda* (c.1700 BC) dedicates instead 114 of its 1028 hymns to another perception enhancing substance, soma. Early archaeological research suggests that soma was a drink originating in Aryan culture and that it contained a mixture of opium, cannabis and ephedra. More recent theories point instead to a bright red mushroom with white speckles, the fly agaric (*Amanita muscaria*), as the likely main ingredient of the soma of antiquity. Its ability to alter experience is mentioned in hymn 48, verse 3 of the *Rig Veda* which states:

We have drunk Soma and become immortal; we have attained the light, the Gods discovered.
Now what may foeman's malice do to harm us? What, O Immortal, mortal man's deception?

In the Greek world, from approximately 1700 BC the cult of Demeter and Persephone organised a sacred procession each September which started in Athens and ended in a mysterious initiation ceremony

opposite left: The heat of the tracks and approaching train are clearly visible here, as well as the sun-warmed metal roofing and glass skylights

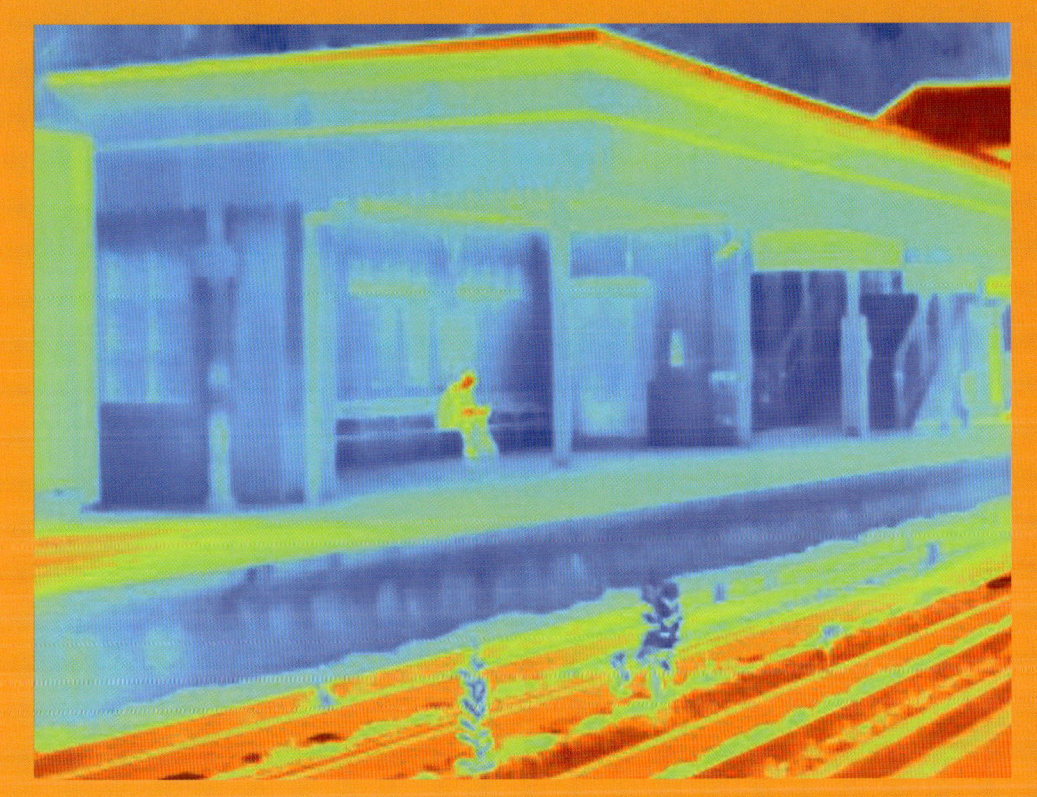

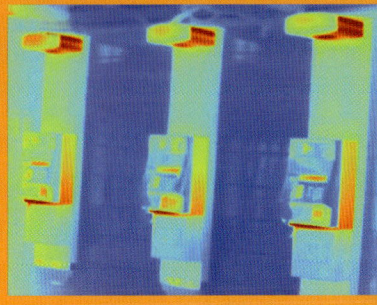

opposite right: Overhead fluorescent lights glow with heat as people exit the station

left: The hot metal of the rail tracks contrasts with the cool plants that grow between them

above: A row of telephone boxes light up an underground tube station

at the Temple at Eleusis. In their 1978 book *The Road to Eleusis: Unveiling the Secret of the Mysteries*, the authors Gordon Wasson, Albert Hofmann and Carl Ruck proposed that the states of spiritual awareness and transcendence which typified the Eleusinian Mysteries were in part produced by a drink, kykeon, which was prepared from the rye-attacking fungus ergot, harvested from the rye plants of the Athenian plain. Ergot contains the naturally occurring lysergic acid, a base product from which the more powerful and better known LSD can be synthesised. The Mysteries, which continued until 396 AD, were experienced by thousands of individuals including such dignitaries as Aristotle, Sophocles, Plato, Aeschylus, Cicero, Pindar and the Roman emperors Hadrian and Marcus Aurelius.

Use of perception altering substances was also widespread in the Americas. The first European observation of the use of such chemicals in the New World occurred during Columbus' second voyage (1493-1496). The Spanish friar Ramón Pané wrote of cohoba sniffing by the Taino Indians of Haiti, who used it to communicate with the spirit world. Cohoba snuff is prepared from the seeds of the yopo tree and contains the active component DMT. Modern archaeology has dated a number of Caribbean *cohoba* pipes to at least 400 BC, suggesting a long history of cohoba usage in the pre-Columbian Americas.

While most pre-Columbian records were destroyed by the Spanish, the mushroom stones from the temple ruins of Guatemala suggest that psilocybin mushrooms have been in use for ritualistic, divination and healing purposes since at least 1000 BC. The Franciscan friar Bernardino de Sahagún (1499-1590) wrote the first Western account of the use of these psilocybin mushrooms in his *Historia general de las cosas de Nueva España*. According to Bernardino de Sahagún these mushrooms were known to the

Aztecs as *teonanácatl*, meaning "flesh of the gods", and were widely used in formal ritual. Bernardino de Sahagún also wrote the first Western account of the use of peyote among the Aztecs of Mexico. The acrid tasting peyote cactus is native to large areas of northern Mexico, and the *Lophophora williamsii* variant contains large amounts of mescaline, an active chemical with perception enhancing powers comparable to LSD. Based on Aztec historical chronology, Bernardino de Sahagún estimated that peyote was used by the Chichimeca and Toltec Indians at least 1890 years before the arrival of Columbus.

Another perception enhancing substance of the Americas which is mentioned in early Spanish records is ololiuqui, a preparation which has been found to be rich in lysergic acid LSA. Modern archaeological and ethnographical research has suggested that ololiuqui was prepared from the seeds of the *Rivea corymbosa*, or white flowered morning glory, a plant which was common in the New World.

The use of bhanga, soma, kykeon, cohoba, teonanácatl, peyote, ololiuqui, and other chemical substances by our ancestors suggests an innate desire to explore and experience the world in a deeper and richer manner. From the time of the earliest human cultures it seems that there has been a constant need to explore and understand the world which has lead to a continuous and exhaustive search for ways of enhancing perception.

UNDERGROUND TRAINS

Perhaps the most fascinating place to experience thermal vision is on and around the trains of the underground, where everything can seem unfamiliar. The trains and tracks glow with energy, electrical components burn bright red as electricity moves through them, and everywhere people emit heat as they go about their daily lives.

below: The lights and electrical components can be seen on the back of this train

below right: The doors open to reveal a warm interior

opposite above: People exit the train. The burning red spots overhead are the station lights as they lose energy through heat

opposite below: This sequence shows the train leaving the station. Thermal vision allows us to see into the darkness of the tunnel, highlighting the tracks and cabling along the walls

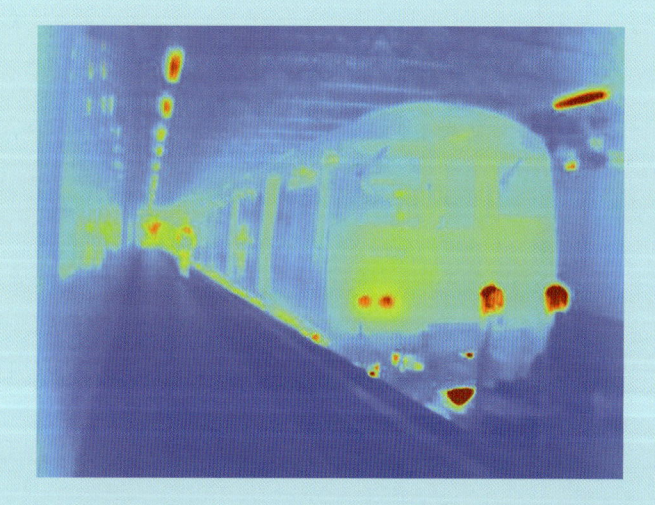

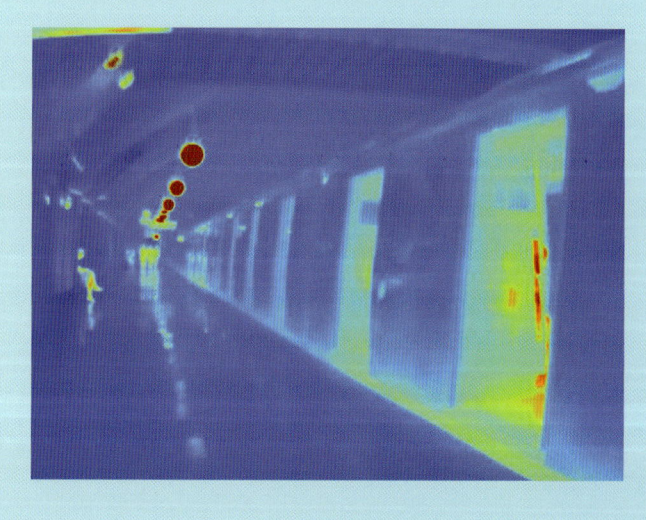

73

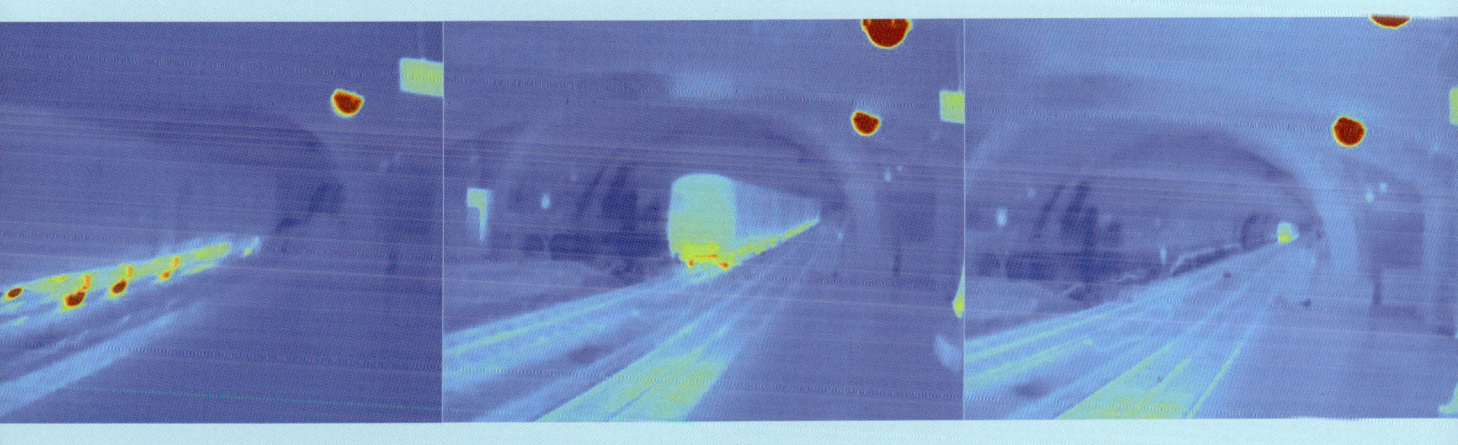

VEHICLES

Vehicles are essentially devices for converting energy into motion, and so they produce distinctive thermal signatures from their engines and moving parts. Through thermal eyes our view of the world returns to the early days of motoring, when road vehicles were little more than an engine and a set of wheels. Thermal vision creates a world where the domain of energy and machinery is distinct and segregated from the domain of living creatures: cars and other vehicles are depersonalised and presented as merely functional objects.

right: The heat from the engine, exhaust system and tyres radiates out from this taxi. The darker spots on the wheels are the brake calipers, hot from the friction of the brake pads against the tyres. The warm area beneath the car may be heat reflecting off the ground

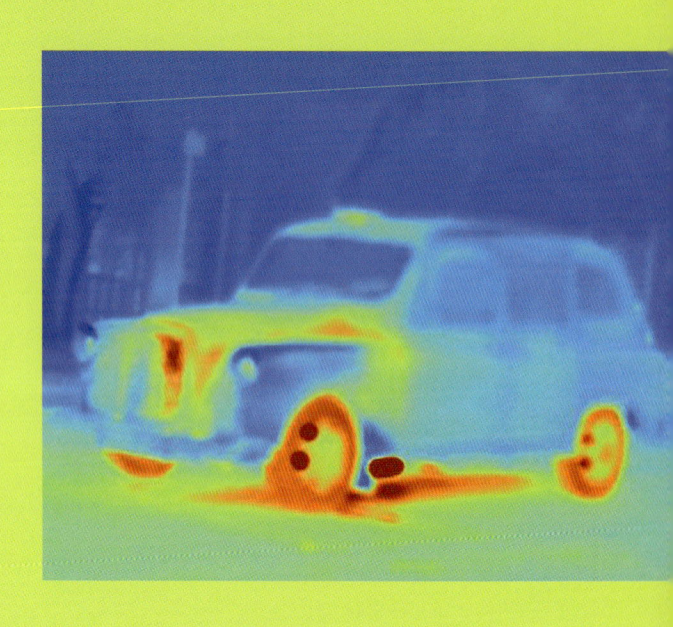

In modern terminology we speak of the active components of these natural substances as drugs, assigning them names such as DMT (cohoba), LSA (kykeon and ololiuqui), LSD, mescaline (peyote), muscimol (soma), muscazone (soma), psilocybin (teonanácatl) and THC (cannabis). What these drugs have in common is the ability to alter human perceptual and cognitive function. With the exception of cannabis, most of these drugs directly alter the activity of the serotonin receptors of the brain. Aldous Huxley, in his 1954 book *The Doors of Perception*, first suggested the concept of the "reducing valve" of the brain. According to Huxley, drugs such as those listed above counteract the brain's ability to selectively filter perceptions, memories and emotions, thus enhancing the focus on the signals from our biological receptors. Huxley believed that the human nervous system had evolved over millions of years in such as way as to filter any perceptions which were not absolutely relevant to survival, simplifying and categorising things, eliminating most of what is present in the raw perceptual stimuli. Perception enhancing chemical substances therefore act to moderate, or even neutralise, the reductive processes of the mind.

The term "psychedelic" was first introduced in 1957 to describe this general class of drugs. The psychiatrist Humphry Osmond derived the term from the Greek words for "mind" and "manifesting", emphasising the ability of these chemicals to reveal both the raw contents of perception and the subsequent elaboration performed by the mind. Personal experimentation with psychedelics accelerated during the 1950s due to the widespread and legal availability of mescalin, LSD and psilocybin. Interest

above left: As the bus travels, friction causes the tyres to warm up

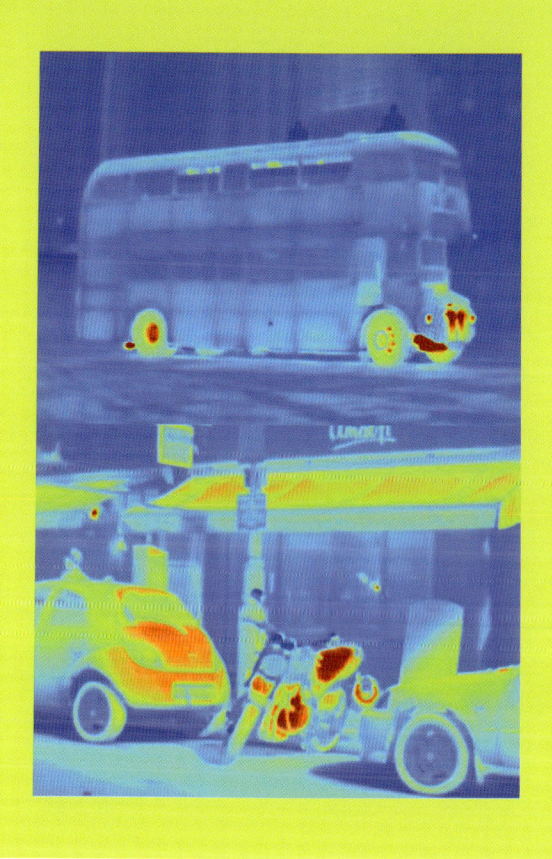

left: The engine on this motorcycle is still hot from its recent use, though the car behind it is warm from the heat of the sun

above: The car's warm engine heats up the front body panel

rapidly spread beyond the confines of the pharmacological and psychological research laboratories to a wider community of people who were searching for a wider range of answers.

After this growth in the 1950s, the 1960s soon developed into the golden era of psychedelic experimentation. A cannabis and LSD fuelled wave of experimentation lead to new psychedelic citizens and new social structures. In the early 1960s the hippie was born, emerging from the culture of the hipsters who moved into San Francisco's Haight-Ashbury district. A youth movement, counterculture and beat generation was established. Communes such as Millbrook, Drop City and Morning Star germinated and grew, geodesic domes were built, and psychedelic purple entered the colour chart. 1960s' psychedelic exploration went far beyond the simple act of enhancing perception and became a lifestyle in itself, producing new mystical and religious movements and pointing the way to new cultures and new worlds. New terms such as "crash", "trip" and "voyage" entered languages across the globe. New perception was taken to mean a new world.

One leader of the new movement was Timothy Leary, a Harvard University psychologist who abandoned his scientific training to pursue a life of psychedelic experimentation. Leary was convinced of the potential of psychedelics to assist the transcendence of man. He is the author of several books, but probably best known for his seminal work *The Psychedelic Experience* (1964), in which he outlines the perceptual and cosmic feasts produced by the use of LSD and other perception enhancing drugs. One passage expresses particularly well the general mood of the time:

CHURCHYARD

In recent years, science has revealed how ultrasound, temperature or other environmental conditions can produce in people those sensations of déjà vu, presence or fear which are behind ancient beliefs in ghosts and spirits. While traditional environmental conditions have led to many reported sightings, what might happen if we all took advantage of thermal vision to enhance our perception? How spooky do these scenes of a country churchyard appear to you?

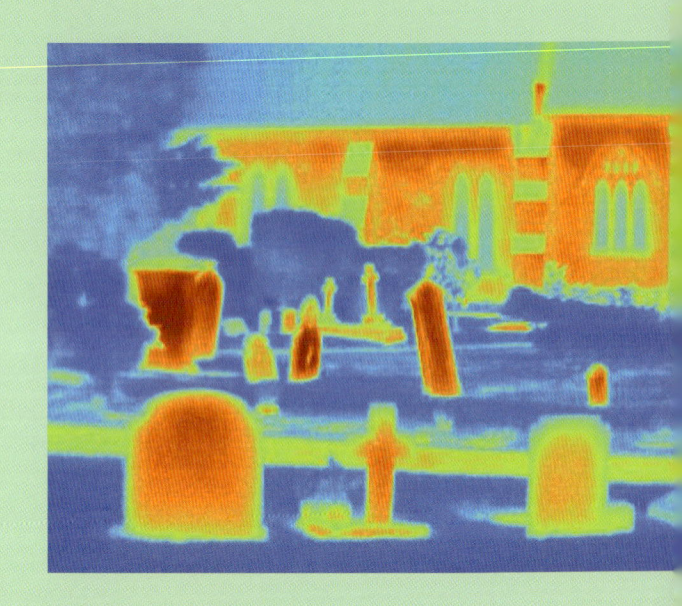

right: The sun heats the headstones and church in this graveyard, but the grass and trees remain cool

A psychedelic experience is a journey to new realms of consciousness. The scope and content of the experience is limitless, but its characteristic features are the transcendence of verbal concepts, of space-time dimensions, and of the ego or identity. Such experiences of enlarged consciousness can occur in a variety of ways: sensory deprivation, yoga exercises, disciplined meditation, religious or aesthetic ecstasies, or spontaneously. Most recently they have become available to anyone through the ingestion of psychedelic drugs such as LSD, psilocybin, mescaline, DMT, etc.

Transcendence was the driving force behind the ritualistic experimentation of ancient societies and was also the driving force behind the intense social change of the psychedelic 60s. The ability of psychedelic substances to not just enhance perception but to also bring it to the very forefront of consciousness seems to have appealed to the need for answers about the "things in themselves". Enjoyment derived from the sensory effects drove the psychedelic train, but is was a desire to transcend which first lead it out from the station. In *The Doors of Perception*, Aldous Huxley provided a description of the feelings of transcendence which are brought about by mescaline:

Mescalin raises all colours to a higher power and makes the percipient aware of innumerable fine shades of difference, to which, at ordinary times, he is completely blind. It would seem that, for Mind at Large, the so-called secondary characters of things are primary. Unlike Locke, it evidently feels that colours are more important, better worth attending to, than masses, positions and dimensions.

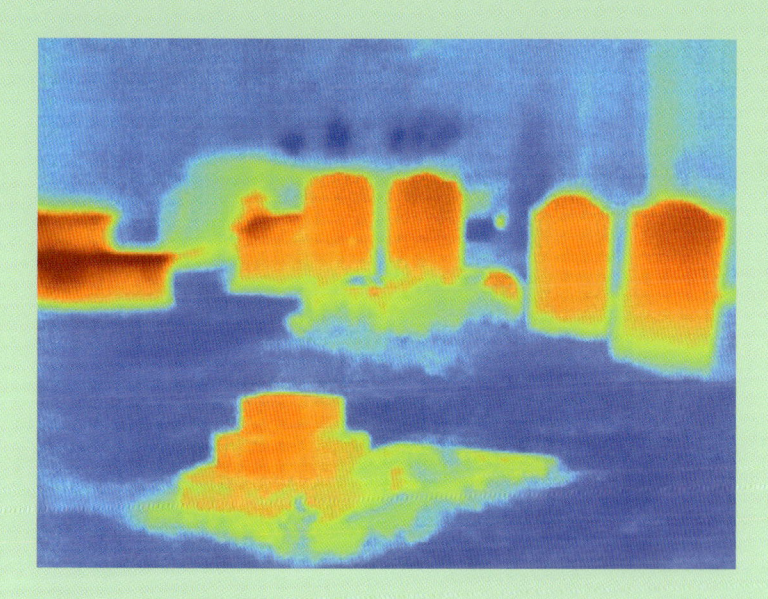

above and above right: The headstones are warmed by the sun, reaching the hottest temperatures at the top, and the lowest at the bottom where they lose heat to the ground

right: The metal railings around this memorial are cooled by the breeze and retain a lower temperature than the stone monument

The overwhelming transcendental force of pure sensation is also documented by numerous episodes reported by Robert Masters and Jean Houston in their influential book *The Varieties of Psychedelic Experience* (1966). In one example, one their more enthusiastic test subjects describes the psychedelic sensations as follows:

Along with all this there were torrents of ideas, some amplifications of my own past thinking, but others that were strange and entered my mind as if from without. At the house, when we returned and the effects were much less, it seemed to me that what I had experienced was essentially, and with few exceptions, the usual content of experience but that, of everything, there was more. This more is what I think must be meant by the 'expansion of consciousness' and I jotted down at that time something of this more I had experienced.

The consciousness-expanding drugs, I wrote then, enable one to sense, think and feel more.

Looking at a thing one sees more of its colour, more of its detail, more of its form.

Touching a thing, one touches more. Hearing a sound, one hears more.

Tasting, one tastes more. Moving, one is more aware of movement. Smelling, one smells more.

The mind is able to contain, at any given moment, more. Within consciousness, more simultaneous mental processes operate without any one of them interfering with the awareness of the others. Awareness has more levels, is many-dimensioned. Awareness is of more shades of meaning contained in words and ideas.

One feels, or responds emotionally with more intensity, more depth, more comprehensiveness.

There is more of time, or within any clock-measured unit of time, vastly more occurs than can under normal conditions.

There is more empathy, more unity with people and things.

There is more insight into oneself, more self-knowledge.

There are more alternatives when a particular problem is considered, more choices available when a particular decision is to be made. There are more ways of 'looking at' a thing, an idea or a person …

STONEHENGE

For centuries people have speculated about the nature and purpose of Stonehenge, with popular theories ranging from religious temple to alien landing site. Thermal vision confirms the well-known warm properties of the bluestones and highlights various cracks and defects in the structure, but unfortunately it does not solve the ancient riddle or reveal "the secrets of the stones". Perhaps another form of perception enhancement might one day reveal deeper and darker secrets.

The bluestones at Stonehenge appear warmer than the surrounding grass and sky as they have a higher thermal conductivity. The site is famous for its stones which align almost perfectly with the sunrise on the summer solstice

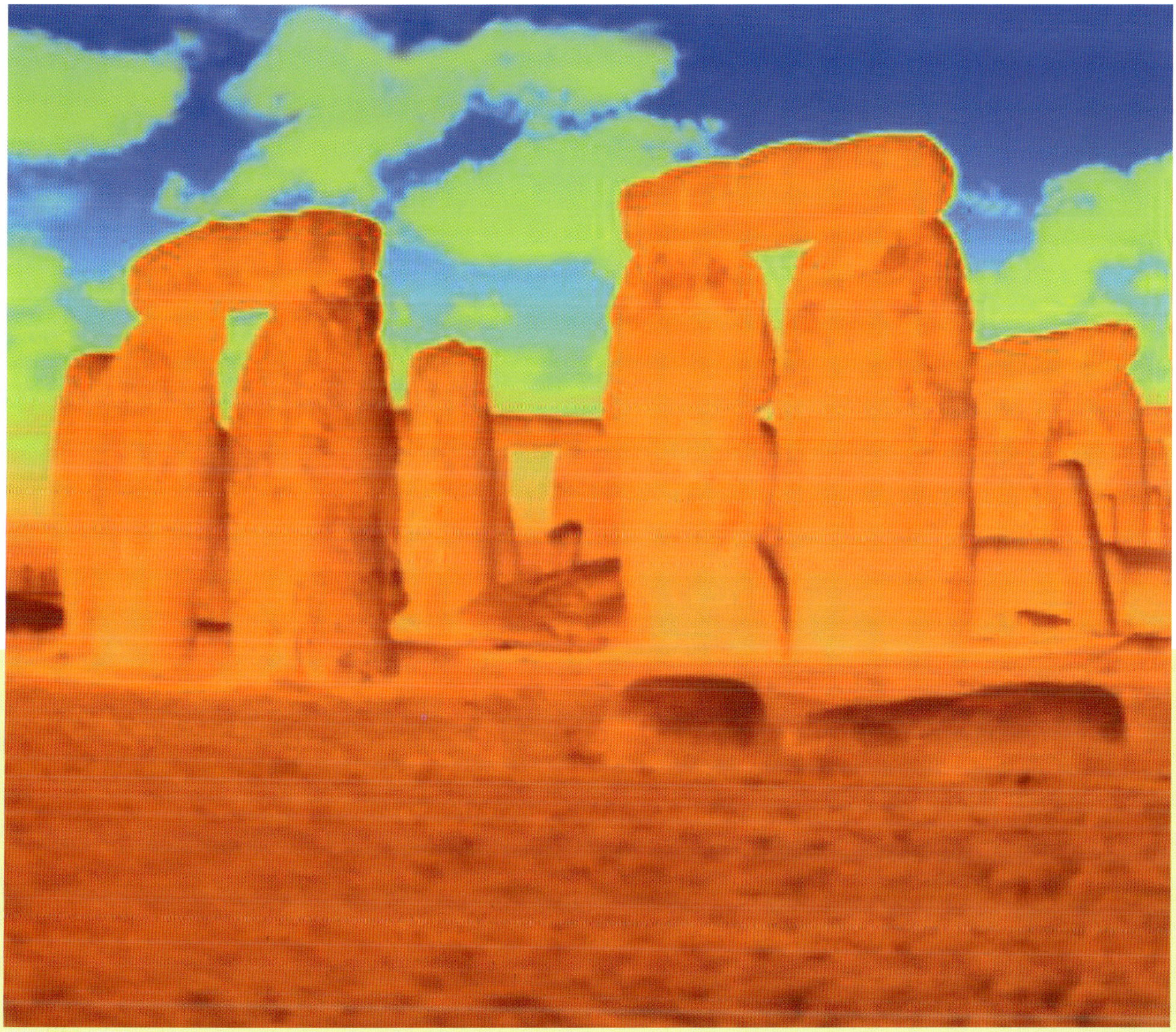

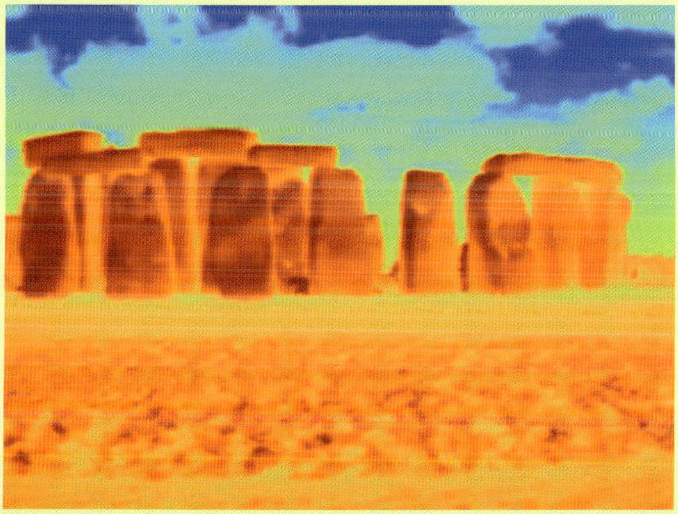

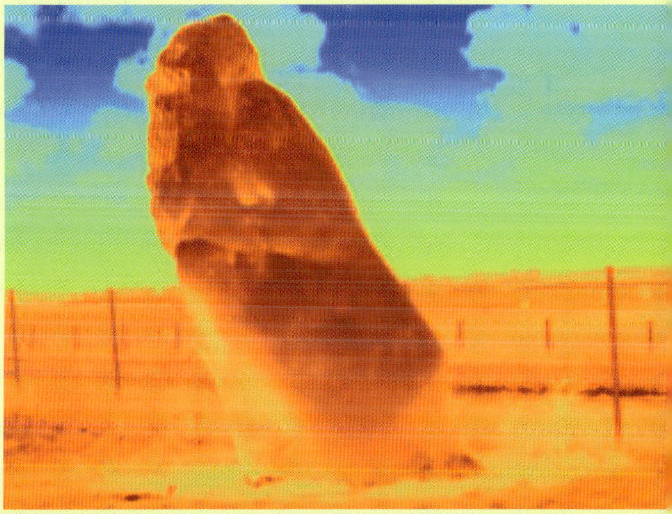

BENCHES AND CHAIRS

Benches and chairs are surprising to look at through thermal eyes. Parts of these essential everyday items tend to be cool on a warm summer's day, and the shadows which they cast can cool the surroundings beneath or behind them. Thermal vision reveals how these objects are crafted with different materials, in different thicknesses, and how this plays a part in the design.

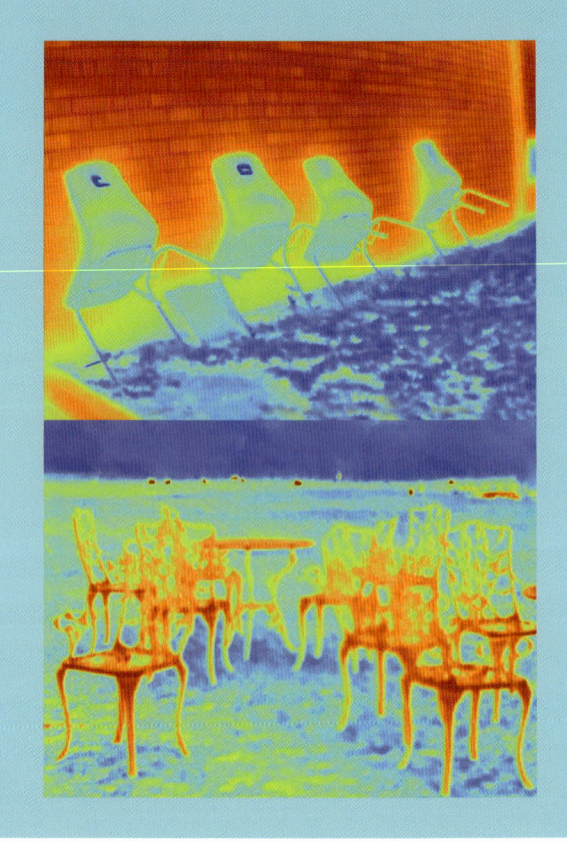

above right: A row of plastic chairs are leaned against a brick wall after cleaning

below right: A set of metal outdoor chairs and tables warmed by the sunlight. The dark red spots in the distance are rabbits

When searching for the causal links between the chemical pharmacology and the resulting psychological and sociological manifestations of transcendence, Masters and Houston noted:

That the subject is seeing "better" or "more clearly" the colors, lines, and other specific properties and parts of things appears to be often explained by the fact that the object in such cases no longer is being apperceived in terms of function, symbolism or label categorizations of the object not accessible to sense perception alone and which usually work to dilute the immediacy of the perception.

Adopting a less scientific language in *The Psychedelic Experience*, Timothy Leary described the same invasive and irresistible force of transcendence:

It comes about this way. The subject's awareness is suddenly invaded by an outside stimulus. His attention is captured, but his old conceptual mind is not functioning. But other sensitivities are engaged. He experiences direct sensation. The raw "is-ness." He sees, not objects, but patterns of light waves. He hears, not "music" or "meaningful" sound, but acoustic waves. He is struck with the sudden revelation that all sensation and perception are based on wave vibrations. That the world around him which heretofore had an illusory solidity, is nothing more than a play of physical waves. That he is involved in a cosmic television show which has no more substantiality than the images on his TV picture tube.

below left: A wood and concrete bench cools its surroundings by casting a shadow on a warm day

below right: A wood and steel bench casts a shadow on this underground station platform, and a lone discarded bottle stands out from the warmth of the seat

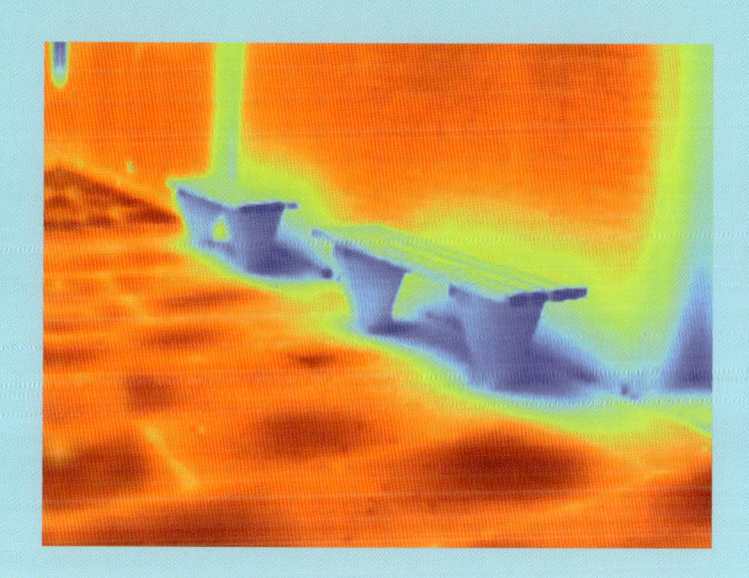

The changes in perception which accompany the ingestion of psychedelic drugs is profound. As the chemical effects begin to take hold, one notes an increase in depth perception and a distinctly greater ability to separate borders and outlines. The changes in serotonin concentration appear to affect the receptor fields of the rods and cones of the eye, of the ciliated cells of the ear and of the mechanoreceptors of the skin, muscle and tendons. Sensitivity to both time and space increases. Vision changes from a world of watercolour prints to one of oil paintings, from the dull and bland to the bright and interesting. Surface textures become fascinating and three dimensional objects such as bubbles become totally mesmerising, particularly if in motion. Objects can bend and melt in response to our attention and focus, literally bending to our personal, directed will. Colours become brighter and more interesting, sounds reveal unexpected texture and complexity, and the world all around suddenly becomes dramatically more compelling. The sound of a spoon falling to the ground can crash into the mind with the force of a catastrophic earthquake. Psychedelic substances seem to bridge the gap between our inner subjective world and the outer objective world.

The dramatic effects of these substances and the cultural changes they brought about via the hippie movements and communes of the 1960s has lead to the creation of a vast amount of literature. Among the many thousands of pages of psychedelic literature, no author seems to have expressed more eloquently the sensations evoked during these enhanced perceptual explorations than Aldous Huxley. In *The Doors of Perception* he fluidly describes the experiences of a single afternoon in 1953 in Hollywood when he first ingested mescaline, the active agent of the peyote cactus. As a gifted writer and poet, Huxley expresses the dramatic alterations in perception perhaps better than anyone. Sections of *The Doors of Perception* express what many people experience, but struggle to describe, in that very limited medium which is language. The following passage is an excerpt from his account of that fateful afternoon:

I continued to look at the flowers, and in their living light I seemed to detect the qualitative equivalent of breathing -but of a breathing without returns to a starting point, with no recurrent ebbs but only a repeated flow from beauty to heightened beauty, from deeper to ever deeper meaning.

Also from that first experience comes the now famous description of the chair:

From the French window I walked out under a kind of pergola covered in part by a climbing rose tree, in part by laths, one inch wide with half an inch of space between them. The sun was shining and the shadows of the laths made a zebra-like pattern on the ground and across the seat and back of a garden chair, which was standing at this end of the pergola. That chair – shall I ever forget it? Where the shadows fell on the canvas upholstery, stripes of a deep but glowing indigo alternated

PEOPLE

Of all the living creatures on this planet, perhaps the most complex and fascinating are human beings. People are everywhere – active, dynamic and vibrant. With much of the colour, camouflage and confusion of the visible spectrum removed, people stand out in front of thermal eyes as forms of pure, raw, existence. Thermal eyes directly see the corporeality and physicality of people, stripping away pretence, to reveal pure presence and action. Posture or motion, activity or rest, being or not being, all are laid bare by the universal signifier that we call heat.

below left: A skateboarder performs stunts

below right: A BMX rider performing stunts

opposite above: Students pose for a group photo. Note the different thermal properties of their clothing, accessories and hairstyles

opposite below left: Tourists stop to eat and rest on the grassy area outside this museum

opposite below right: A young couple hold hands, walking on a cool stone surface

83

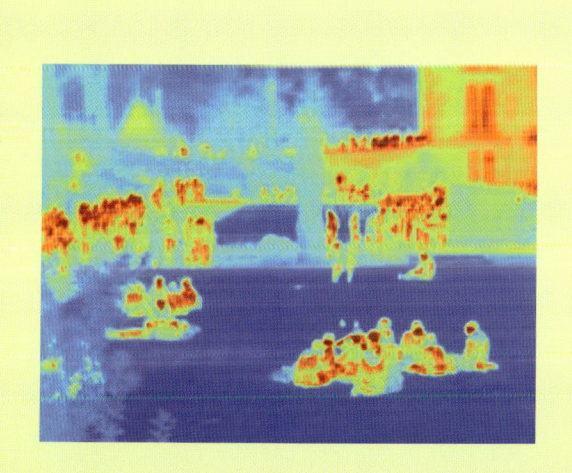

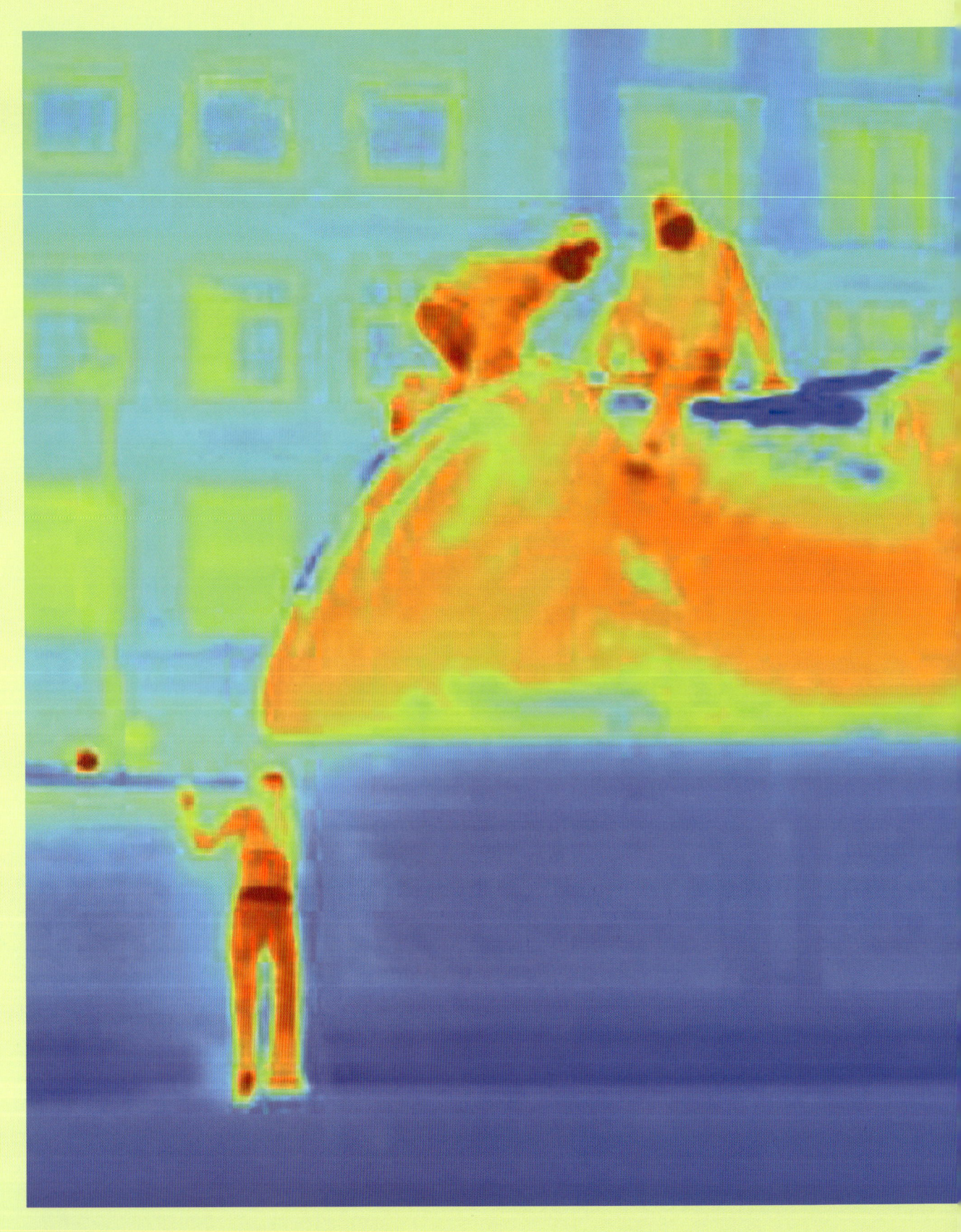

Tourists entertain themselves on the lion statues of London's Trafalgar Square. The bronze lion is warmed by the sun, whilst the granite plinth beneath it remains cool

85

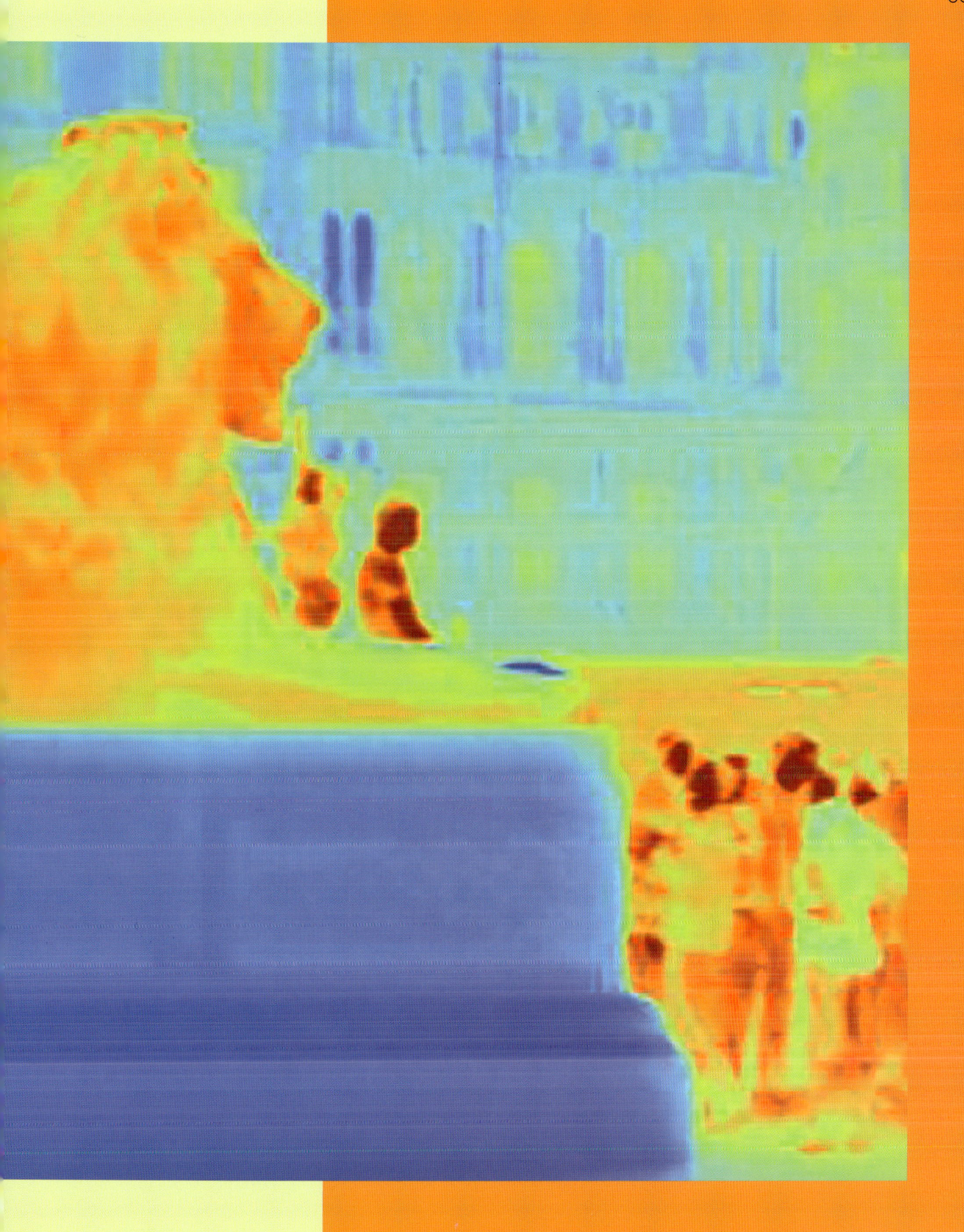

HAIR DRYER

The simple act of drying one's hair becomes an alien experience to someone equipped with thermal vision. Thermal eyes see a blending of person and machine, a whirlwind of shapes and forms which delight the imagination. The scene changes from instant to instant as the heat flows and different areas warm and cool.

below: A hair dryer and brush, warmed by use

right: With thermal vision, one can see the swirling flow of heat transfer from the hair dryer to the brush and the hair

opposite: The thickness of the towel prevents body heat transferring as quickly, and so the towel appears cooler against the woman's body

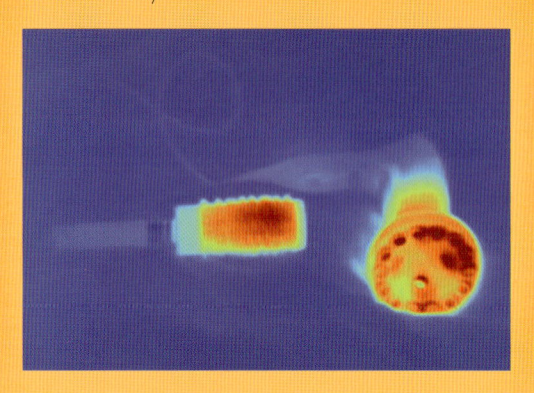

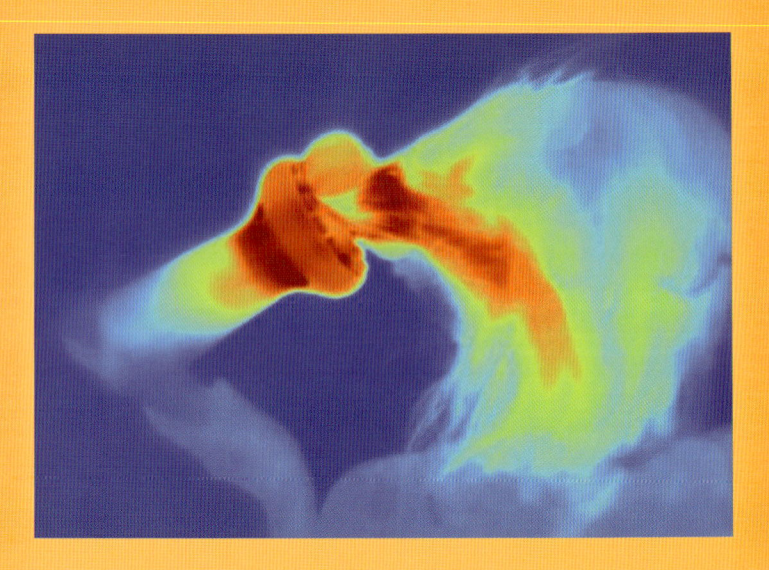

with stripes of an incandescence so intensely bright that it was hard to believe that they could be made of anything but blue fire. For what seemed an immensely long time I gazed without knowing, even without wishing to know, what it was that confronted me. At any other time I would have seen a chair barred with alternate light and shade. Today the percept had swallowed up the concept. I was so completely absorbed in looking, so thunderstruck by what I actually saw, that I could not be aware of anything else. Garden furniture, laths, sunlight, shadow – these were no more than names and notions, mere verbalizations, for utilitarian or scientific purposes, after the event. The event was this succession of azure furnace doors separated by gulfs of unfathomable gentian. It was inexpressibly wonderful, wonderful to the point, almost, of being terrifying. And suddenly I had an inkling of what it must feel like to be mad.

For many, the grip of such experiences proved strong. Albert Hofmann, the Swiss chemist who first synthesised LSD, spent a lifetime investigating both the sources and effects of LSA and LSD; Timothy Leary abandoned science for a lifetime of transcendental and psychedelic exploration; Aldous Huxley became a major exponent of the therapeutic, medical and transcendental benefits of psychedelics and Robert Gordon Wasson spent a large portion of his life searching for psilocybin mushrooms and studying their effects. These individuals, and many more, dedicated a lifetime of work to the pursuit of psychedelics-induced understanding. Besides the best known figures, the pages of websites such as Erowid, Lycaeum, MAPS and many other small organisations which work to provide information about psychedelics explode with the personal accounts of individual psychonauts. Books, magazines and the internet are

87

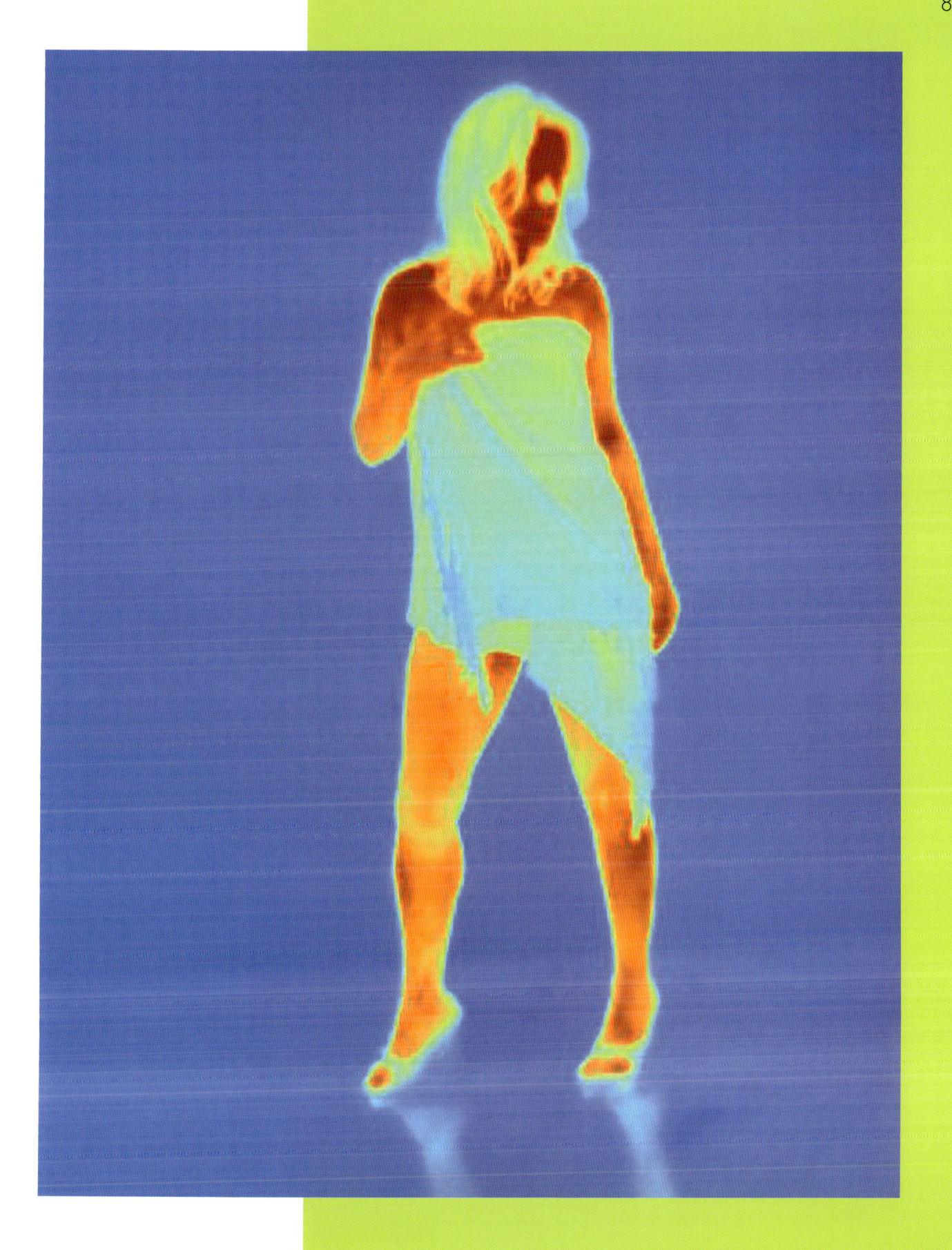

88

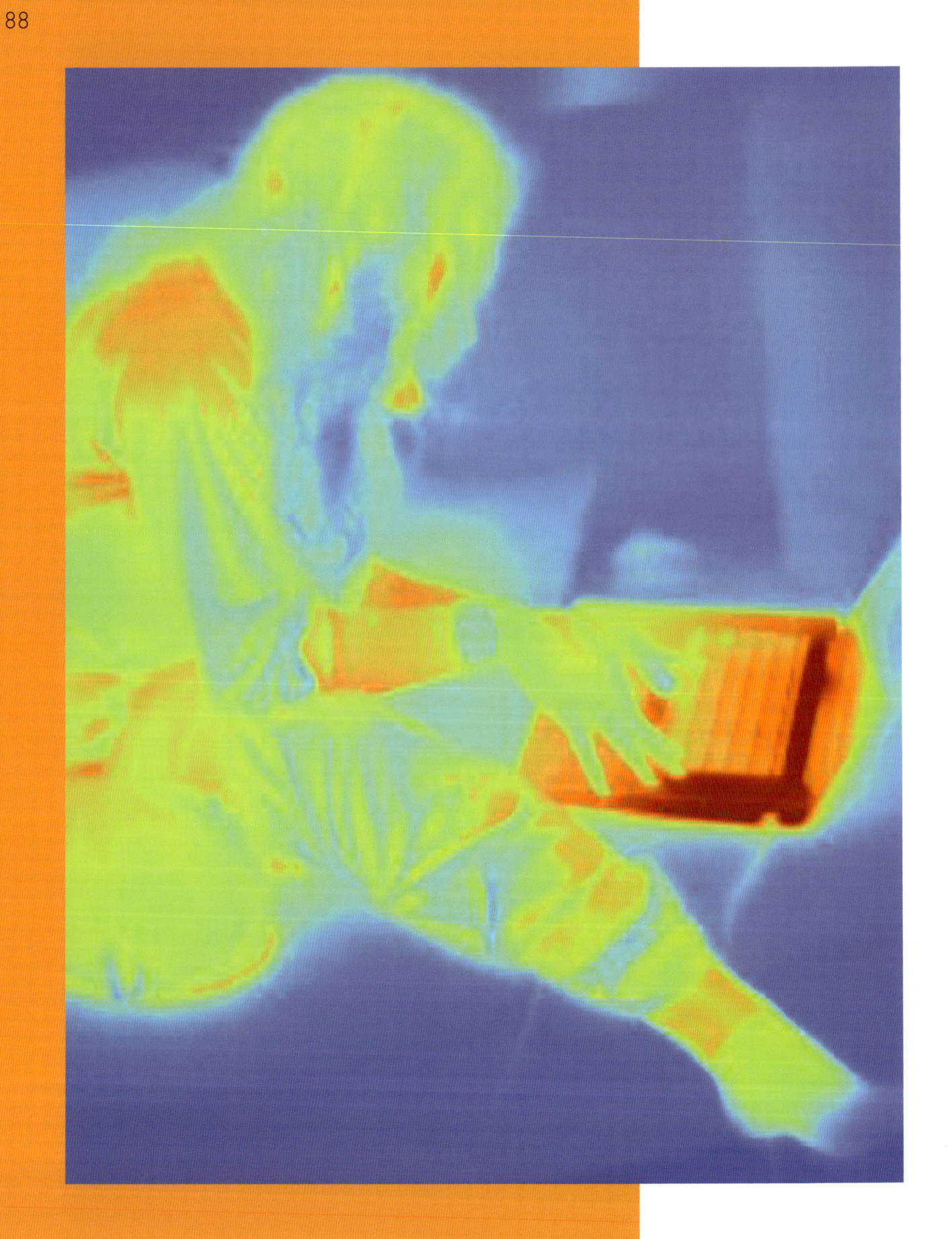

GLASSES

The cleaning of glasses in a pub becomes a ghostly experience when viewed through thermal eyes. The warm glasses contrast strikingly with the cool environment behind the bar, revealing the simple heart of the pub. Enhanced perception uncovers a world which is vastly different from our everyday experience, one in which glass is opaque, cloth is transparent and opaque surfaces reflect.

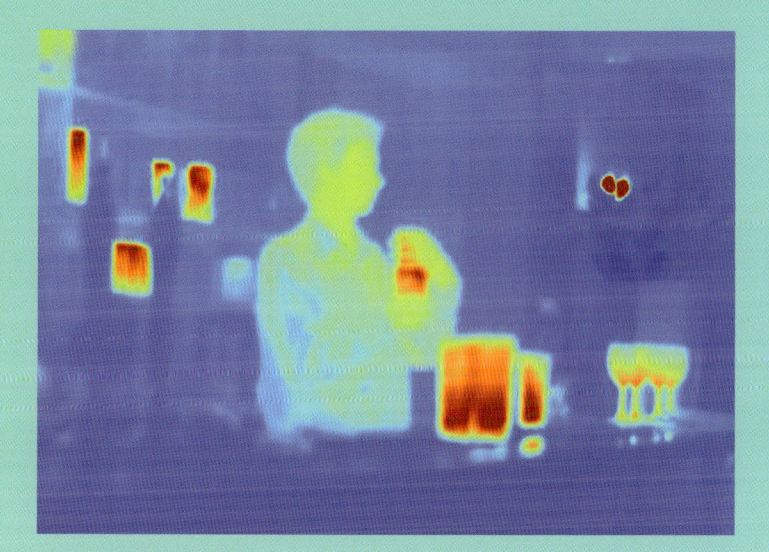

opposite: The outer case of this laptop is warmed by the hot air which is pushed out by the cooling fan

left and below: The warm glasses, straight from the dishwasher, stand out against the cool environment of the pub

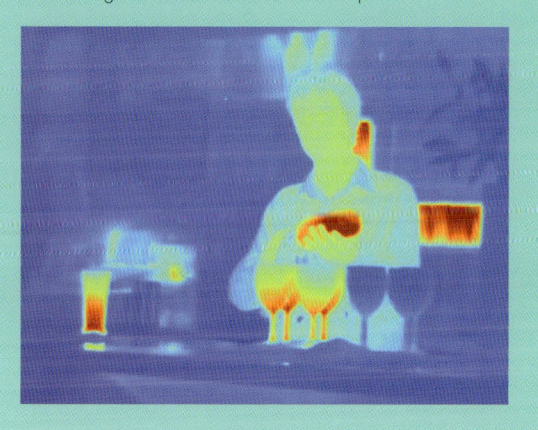

strewn with the psychedelic diaries of thousands of individuals who have experienced some form of deeper understanding though the use of these substances. The motivations behind individual experiments in psychedelia vary, as do the background and professional profiles of the individuals involved. In *Utopiates: The Use and Users of LSD-25* (1964), Dr Richard Blum and his associates noted that

The movement is composed of people who have taken LSD and/or other hallucinogens and see in these drugs a tool for bringing about changes which they deem desirable. The emphasis is on the enhancement of inner experience and on the development of hidden personal resources. It is an optimistic doctrine, for it holds that there are power and greatness concealed within everyone. It is an intellectual doctrine, for it values experience and understanding more than action and visible change. It concerns itself with areas dear to the thinker: art, philosophy, religion, and the nature and potentials of man. It is a mystical doctrine, for it prizes illumination and a unified world view with meaning beyond that drawn from empirical reality. It is a realistic doctrine as well, for it counsels compromise and accommodation between the inner and outer worlds. 'Play the game,' it advises, 'don't let the Pied Piper lead you out of town.' And it is, explicitly, a revolutionary doctrine, although the revolution it proposes is internal, psychological, and by no means novel. It calls for freedom from internal constraints, freedom to explore oneself and the cosmos, and freedom to use LSD and other drugs as the means thereto.

While unjustified fears eventually produced a backlash of public opinion which manifested itself in the anti-drug laws of the late 1960s and early 1970s, the human need for transcendence and for more information about the "things in themselves" has remained.

90

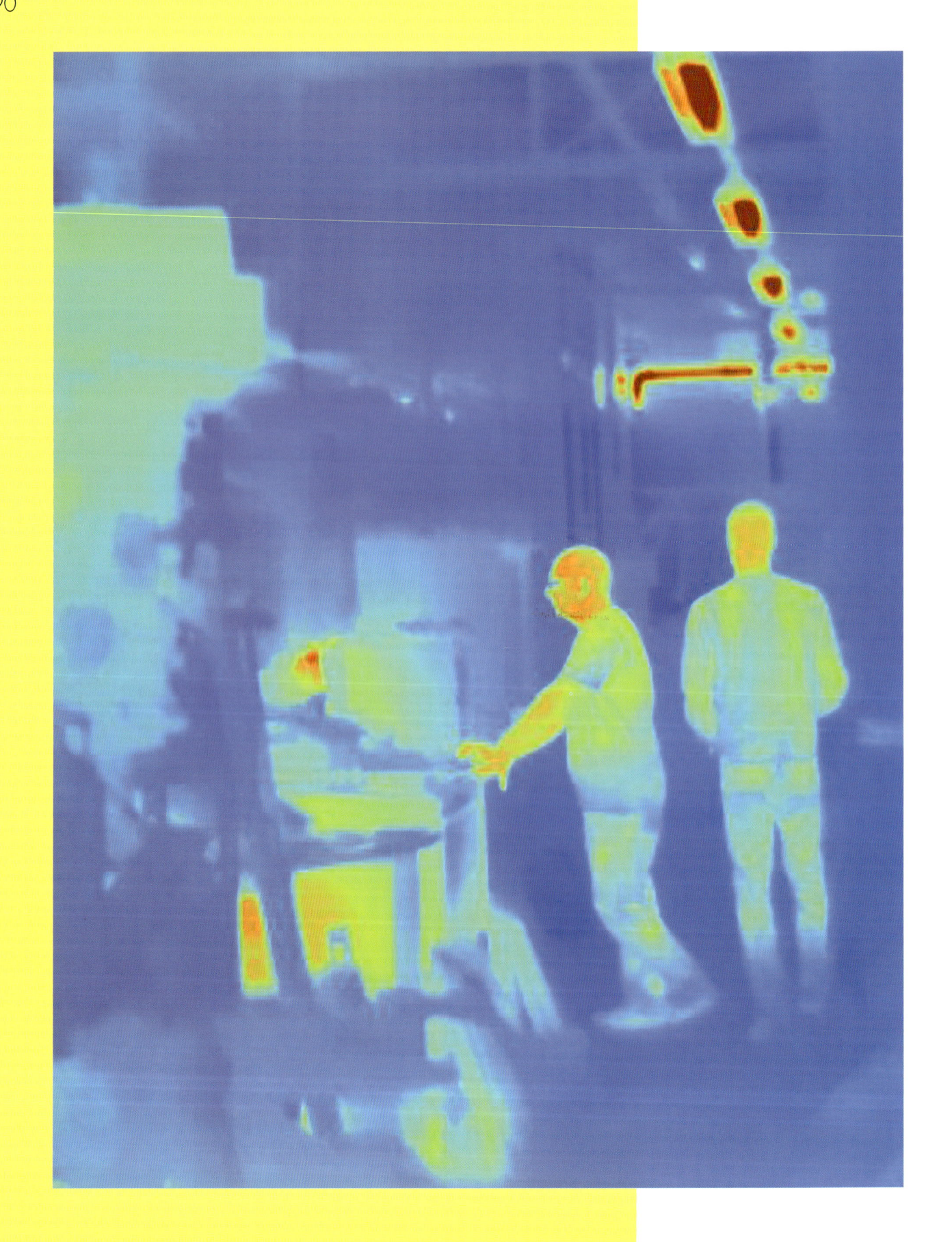

As we venture into the 21st century, the methods used for enhancing perception have entered a new era, one in which technology, rather than rational thought or chemical function, provides the means of transcendence. The now omnipresent availability of technology in its multiple forms provides opportunities for exploring the nature of things in in many different ways. The 21st century is considered by many futurologists to be the "century of the human mind", so it is not surprising that for the first time in history we have available a multitude of powerful technological psychedelics, each capable of rendering visible, audible, touchable, smellable or tasteable any number of scientifically measurable quantities of the world around us. What can be measured can now be perceived, thresholds can be raised or lowered, qualia can be mixed and matched, single or synaesthetic channels of perception can be used and what can be transferred or transformed can now also be perceived. The sky is the limit.

Virtual reality and computer generated worlds are examples of such new technology. The current upwards trend in the use of such mind-expanding scenarios in computer gaming and multimedia suggests that the psychonaut of the 1960s may be morphing into the technonaut of the 21st century. Virtual worlds and virtual lives are rapidly challenging our traditional views of ourselves, of our society, and of our world, breaking down barriers and fuelling social change in a manner not dissimilar to the use of psychedelics in the 1960s. Whilst on the surface it may seem a much quieter revolution, occurring within the home rather that outside it, the commonalities between the two movements have been noted. In the third edition of *Psychedelics Encyclopedia* (1992), author Peter Stafford writes:

Although the affinity between psychedelics and consciousness-changing "brain machines" is fairly obvious, the factors driving so many psychedelic enthusiasts' fascination with virtual reality may be less easily discerned. When asked to speculate why developments in virtual-reality technology so consistently rivet the attention of psychedelically oriented people, several prominent figures present at the Bridge Conference suggested a resonance between the "multiple realities" experienced in psychedelic states and the "multiple realities" that are expected to become available to the skilled pilot of virtual-reality equipment sometime in the near future.

Many psychedelic voyagers have expressed the belief that the realities revealed to psychedelically enhanced perception are to a greater or lesser extent the creation of the user's consciousness – and by extension, so may be the realities evident to "consensus" or "everyday" awareness. Virtual-reality technology allows its pilot to bathe the sensorium in a pool of information designed according to his or her own specifications, thereby providing "a way to experience this alteration of reality in a much more organised fashion" as Bruce Eisner put it.

Stafford further presses this point, stating:

Virtual reality is a manifestation of the increased blurring of the distinction between "solid reality" and the stuff of dreams, thoughts, and the mind. The malleable, rapidly transmuting world into which we're moving is one in which acidheads should have a distinct type of advantage, having already experienced the plasticity and variability of the realities of the mind.

opposite: The machinery and fluorescent lights overhead are easy to spot in this workshop

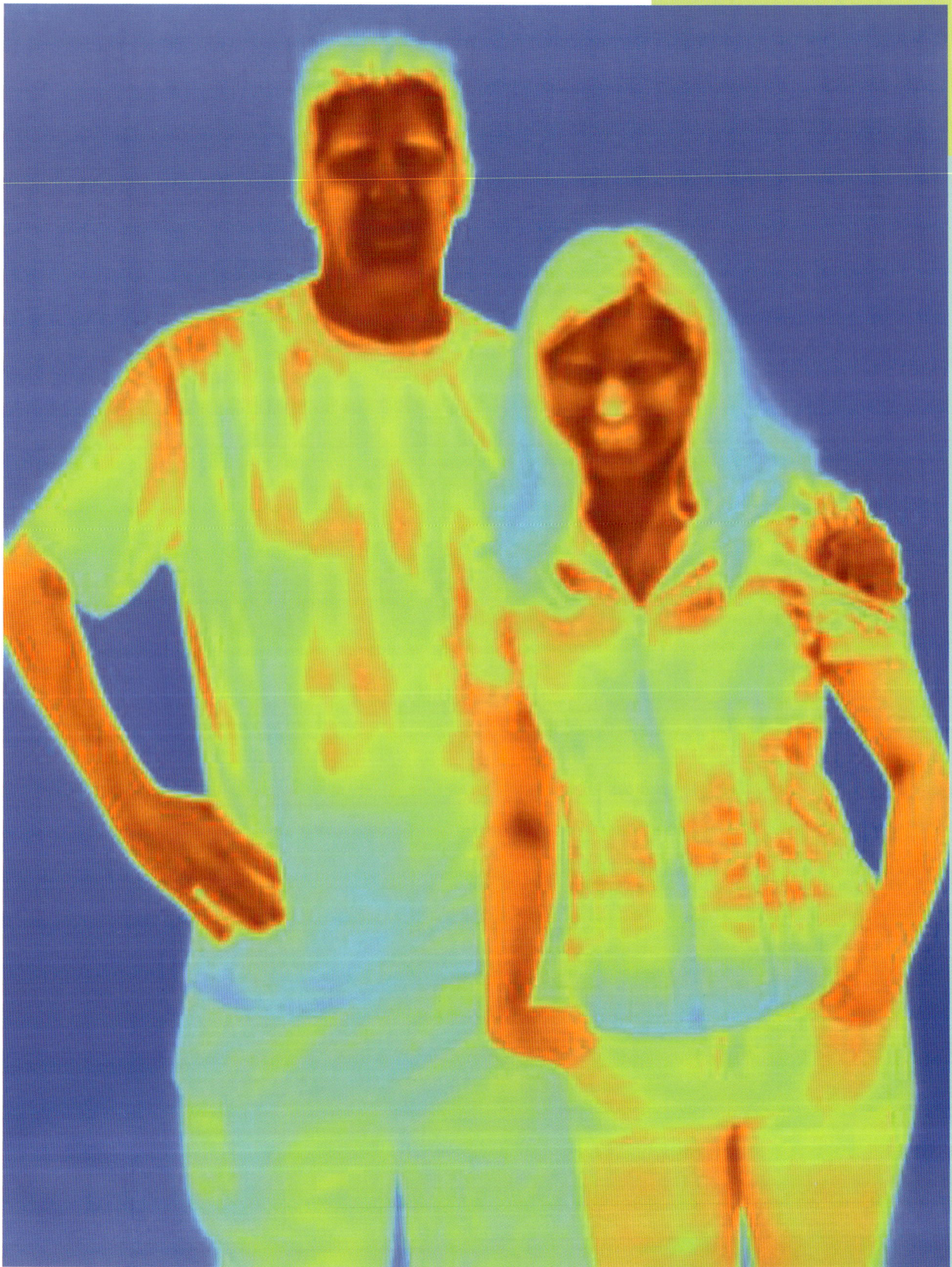

PERCEPTION ENHANCEMENT

So what is meant by the phrase "perception enhancement"? This particular choice of words is probably unfamiliar even to those relatively initiated from fields such as engineering, ergonomics, medicine and psychology. Put simply, what I mean by perception enhancement is ways of exploring the world in which our sensory perception is modified to reveal truths which can be put to practical application. Perception enhancement is the augmentation or transformation of biological sensory abilities so as to differently explore the "things in themselves" which surround us in the physical world.

Upon first consideration, such a concept does not seem particularly contentious and the types of technologies which might be considered perception-enhancing do not appear difficult to list. However, the concept is deceptively complex, with many logical tangles drifting just below the surface. For example, is a telephone which carries voice and sound to distant places to be considered a perception enhancement device? Certainly it makes perception possible at a distance, but should it be classified as a form of enhanced perception? And are binoculars a form of perception enhancement, or only a means for perception at a distance? And what about hearing aids? If humans have the tendency, perhaps even the need, to file things away in tidy conceptual categories, then what should the category of perception enhancement contain?

In ancient times, Roman semaphores, such as the lighthouses of the Saxon Shore, conveyed messages over great distances; a concept which is neither new nor complex. The industrial revolution later brought many other forms of transmission such as the telegraph, telephone, radio and television, but should conveying information be considered a form of perception

opposite: A young couple pose for the camera. Note that the woman's hands are still visible in her pockets

below: A crowd of people walk through an underpass. The combination of the irregular concrete roof and heat reflection from the people beneath it creates a strange effect on the ceiling

enhancement? I would argue against this. From jungle drums to satellite communications, such devices convey sensory information in more or less its original form. Such technologies can be thought of as communication systems rather than perception enhancement systems, transmission rather than transformation. Whilst radio and television are more complex, due to the ability to choose where the original stimuli are recorded and how they are manipulated before transmission, they can still be classed as transmitting rather than transforming information.

Perception enhancement is perhaps best found through technology that selectively manipulates sensory information. Looking back in time, it seems that the possibility of enhancing sensory perception was noted and exploited long ago. We know from archaeological discoveries that optical lenses were used to enlarge or reduce images from at least Greek and Roman times. In the 1st century AD Seneca wrote of using a globe filled with water to magnify written text in order to read it more easily. Vision, therefore, is a sensory channel which has been enhanced in some ways, in some situations, since long ago.

Like Seneca's reading lens, many 21st century products perform selective manipulation of sensory stimuli. Scaling, for example, is now a basic feature of everyday life. Who among us has never turned a radio up or down beyond the natural sound of the live orchestra? And who has never adjusted colour or

below: The horse on the left is wearing a blanket, so we see less of his body heat than the horse on the right

opposite: While the short length of the horse's coat makes him a bright star among the animals of the countryside, the tail disappears

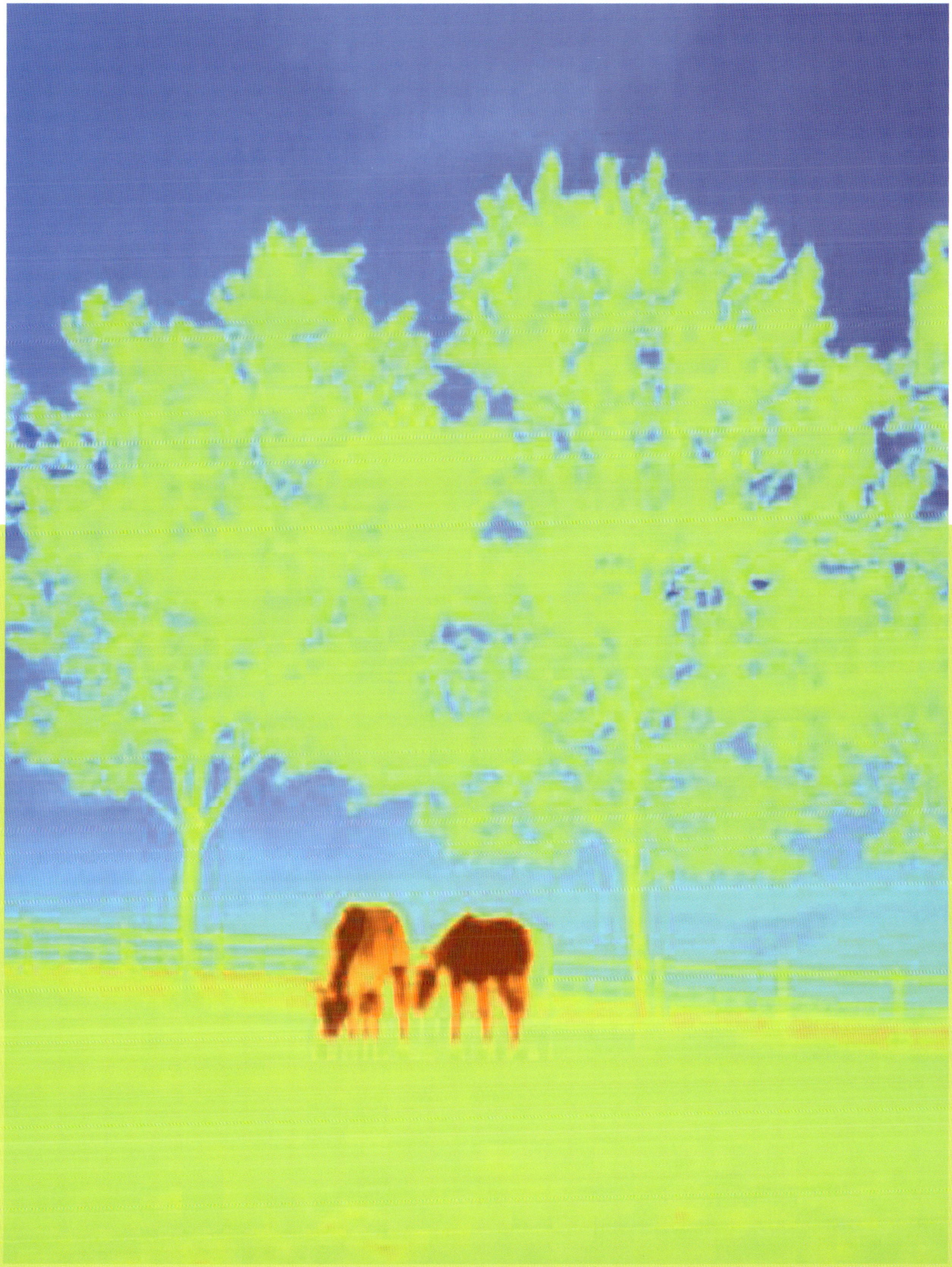

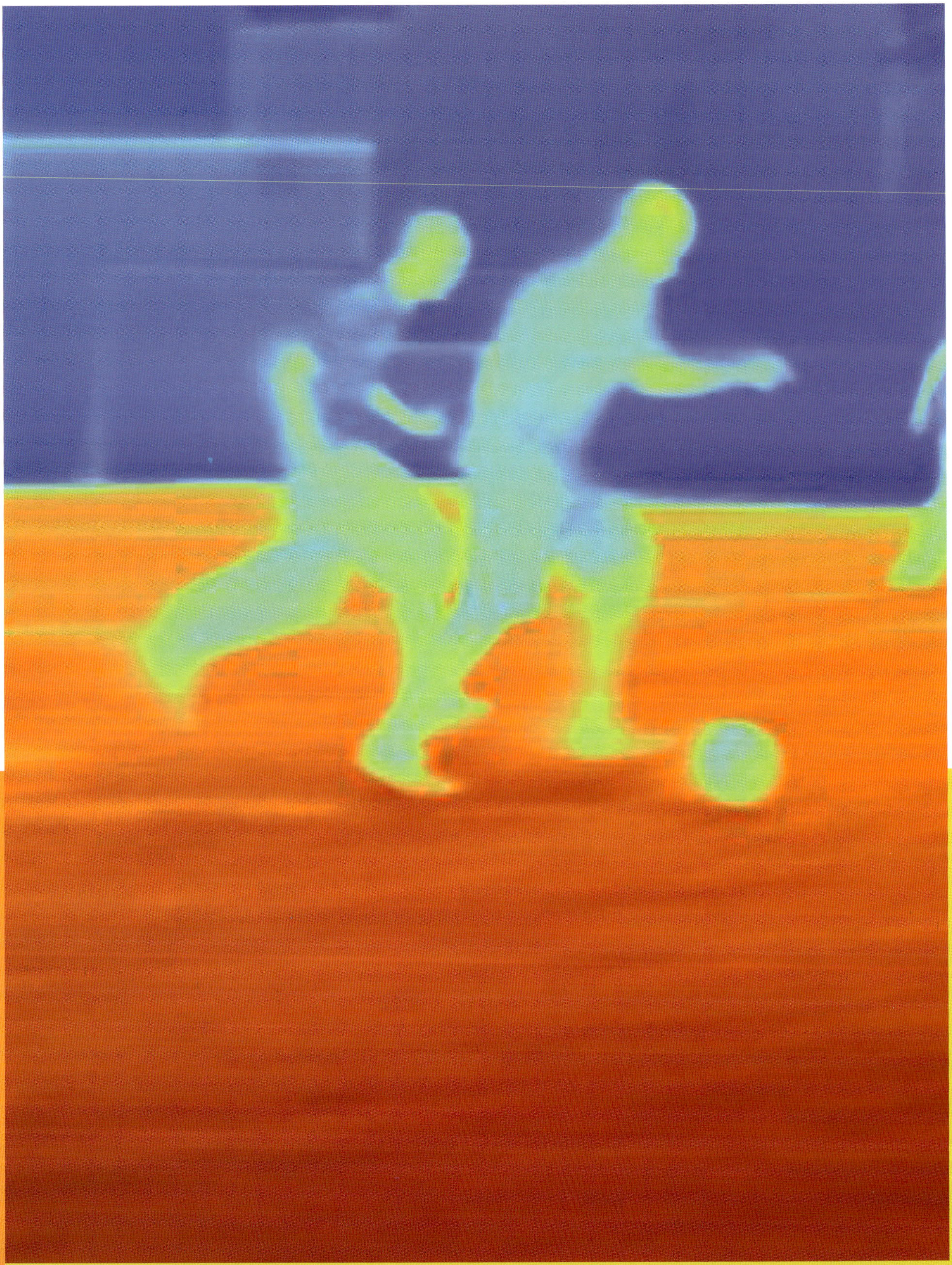

contrast on a television so as to achieve stimuli which are more desirable than the original, natural, phenomena? Scaling is now part of our everyday routine and feels every bit as natural as the original visual, acoustic, tactile, olfactory or gustative sensations produced by the objects around us. With the simple operation of scaling, I believe that we can perhaps begin to speak of perception enhancement.

Looking beyond the ability to scale up or down, many recent technologies also provide sophisticated ways of adjusting the contents of stimuli so as to make them more obvious and intuitive. For example, many modern hearing aids can selectively amplify sound by increasing the volume of only those frequencies which are typical of human speech, reducing the volume of the frequencies of background noise. Other modern hearing aids perform a different trick, automatically steering the axis of greatest microphone sensitivity in the direction of the person who is speaking. In effect, these hearing aids perform the acoustic equivalent of pointing a pair of binoculars at the speaker. Modern hearing aids are full of electronic wizardry which selectively manipulates sound in beneficial ways. Therefore I would argue that they are examples of perception enhancement devices, as they work on only a single type of stimulus (sound) and manipulate only certain of its characteristics. So in its most basic form, a perception enhancement system is any means of selectively manipulating a single sensory stream so as to make it more obvious and intuitive.

Looking beyond enhancements which are performed within a single modality, 21st century technologies also include devices which can measure things which our biological sensory systems cannot. Who among us has not seen an X-ray or thermal image of someone or something in a Hollywood movie? Thermal imaging makes heat visible, permitting us to perceive the temperature of things without actually touching them, as our biological equipment normally requires. And what about virtual reality and virtual worlds? The virtual explorations of the 1980s and 1990s have produced sophisticated ways of expanding understanding through creative encounters between computers and people. Virtual worlds provide infinite possibilities for taking one sensation and transforming it to something altogether different.

opposite: A game of football on a school field. The pitch is mostly soil with very little grass and so heats up directly in the sunlight

right: Groups of people stand in front of the Albert Memorial in London

Heat can be seen, force can be heard, anything which can be scientifically measured can be changed to something else. Devices such as thermal imagers and virtual reality systems pull the great perceptual curtain to one side, revealing behind it a strange new perceptual stage in which the things occupy unexpected places and exhibit unexpected shapes and colours. New technologies are revamping our perceptual theatre, replacing the drab old stage of yesteryear with a shiny new technicolour show.

The ability of 21st century technologies to perform amplification, reduction, bandwidth manipulation and many other selective transformations is changing our world view. While transmission does little to reshape our understanding of the "things in themselves", transformation changes everything. Sensation can be switched from one modality to another, it can be amplified, it can be reduced, it can be reshaped in an infinite number of ways so as to fit our desire to explore and understand. Like a photographer playing with lighting and camera settings, we can now deploy technologies which reconfigure perception so as to reveal the hidden and unexplored. Perception enhancing technologies are helping to convey into inner subjectivity many key aspects of the outer objective world.

In sharp contrast to the loud and noisy revolution of the psychedelic 60s, the current technology-oiled expansion of the human mind is quieter and deeper. There has been much talk in the media about the potentially depersonalising effects of some new technologies, and, conversely, about the social explosion brought about by others. However, academic dissection of the whispering revolution has so far mostly neglected perception enhancing technologies. There is still currently little understanding of what technical professionals are achieving by means of the new technologies and even less awareness of what the creative types are getting up to with the same kit. At times it can feel as if the revolution will all be over before the first shot is heard.

Despite this silence, perception enhancement is already all around us and is destined to become even more ubiquitous as we continue our journey. Like Galileo's telescope, clever 21st century gadgets are

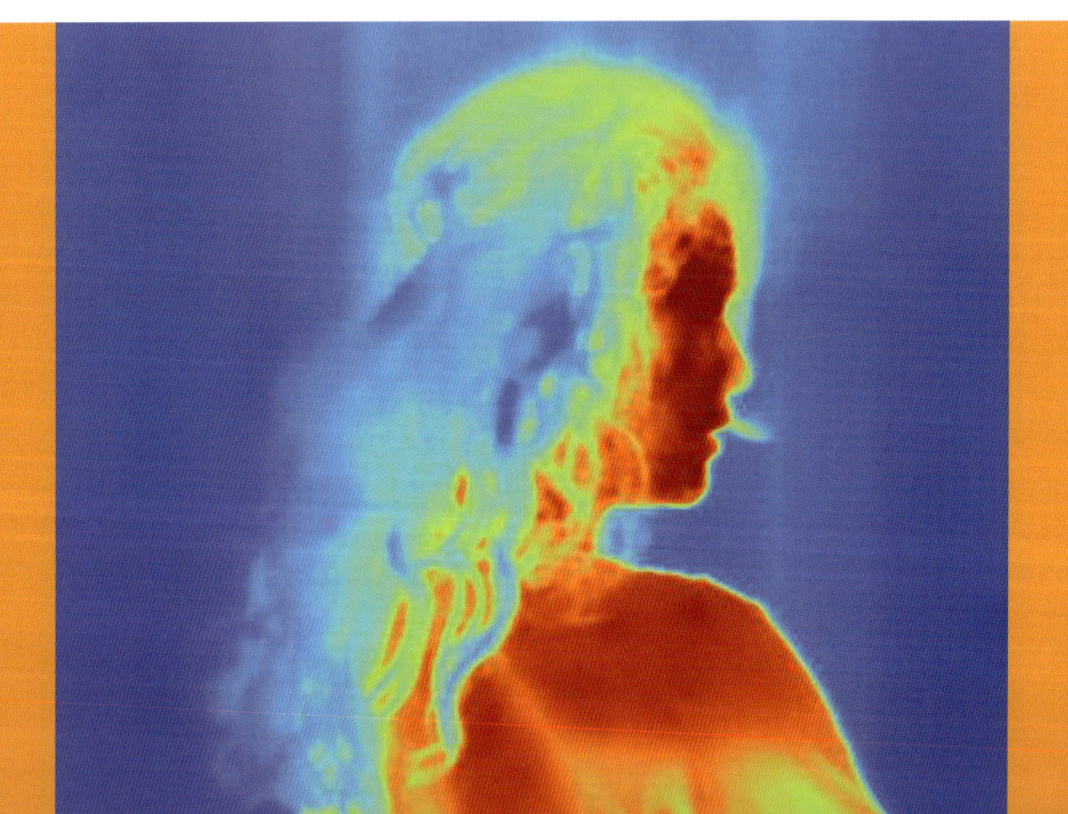

left: Thermal imaging presents a kaleidoscope of colours in this woman's hair

opposite left: The heat from the wheels and cabin of the bus and pedestrians nearby contrasts with the cool trees and road surface

opposite right: A flow of people descend the steps of this shopping centre

opening new worlds for exploration at an ever faster rate. Temperature can now be seen, force can be heard, movement can be tasted, and many other interesting things can be measured, modified and presented to people in a routine manner. Single or multiple synaesthetic channels of perception are now available for use in exploration, learning and enjoyment. The gap between the objective external world and the subjective internal world is being reshaped, reconfigured and remoulded. The combination of the ancient search for transcendence and the modern availability of mind-expanding tools is taking us on a new voyage of discovery.

If we stop for a moment and step back to survey developments from a distance, it seems as though the growing success of perception enhancing systems is somewhat unavoidable. Quite simply, such technologies make life better. Transcending our relatively limited biological sensory systems can bring real advantages. Experiencing the world in different and insightful ways helps us to better understand, better survey and better use what is around us. Throughout history people have attempted new ways of perceiving the world and milestones such as the invention of the telescope or the psychedelic experiences of the 1960s have all brought useful new insights and understanding. Perception enhancing technology can be thought of as an integral part of a continuing 21st century voyage of discovery in which the relationship between people and the planet becomes ever more intimate.

One particular proposal I wish to put forward by means of this book is the use of perception enhancing technology by all those professionals who create things for use by people. Seeing temperature, feeling sound or hearing force can bring interesting new insights to those working in the creative industries. Unlike the 1960s world of inner exploration, the 21st century world is one of outer exploration. New technologies are making it possible to explore the alien surroundings which we call home, stimulating fresh new ideas about the "things in themselves".

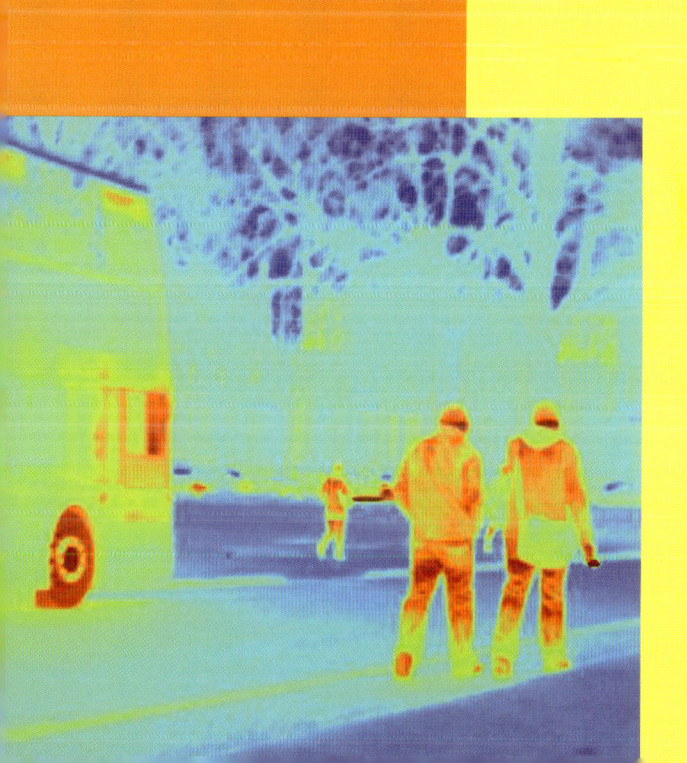

For purposes of illustration this book has focused on the use of one specific perception enhancement technology: thermal imaging. The book provides a small collection of images from enhanced perceptual explorations of the city of London and its nearby surroundings. Some of the images reveal the hidden beauty of heat and light, while others drive home messages about otherwise abstract concepts such as the nature of warm blooded creatures or the problem of global warming. The images reveal that objects are characterised by uneven distributions of heat and that the heat is in continuous movement from place to place. Some images present what is expected, while others surprise, but all are beautiful examples of how the world can look when explored with eyes different from our own.

Thermal imaging was not a difficult choice for this book because in recent years it has become a relatively common and well understood technology. It has many current applications, and is used especially by fire-fighters when searching out smouldering fires, by police when chasing suspects through brush and by military personnel when reconnoitering well dug-in and camouflaged enemies. Thermal imagers have been largely developed and used by engineers, but the technology is rapidly becoming popular in the creative industry too. The cost of thermal imaging cameras has dropped in recent years, the pixel resolution has increased and they have become both portable and easy to use. As a perception enhancing technology, thermal imaging is beginning to come into its own.

The titles of the thermal imaging sections of this book suggest a few of the insights which thermal imaging can help to develop. Electricity, fire, heat, lighting, materials and motion are all concepts which are powerfully revealed through thermal imagining. While modern scientific method provides a multitude of means for determining and graphically presenting the properties of the things, the level of engagement, interaction and understanding achieved is only a shadow of the deep and visceral understanding gained through direct sensory perception. It is one thing to know that a window is losing heat from the home, but an entirely different level of understanding to actually see the heat drifting away.

left: The cool blue windows suggest that this house is not losing much heat

opposite: Heat escapes from the middle window in this block of flats

101

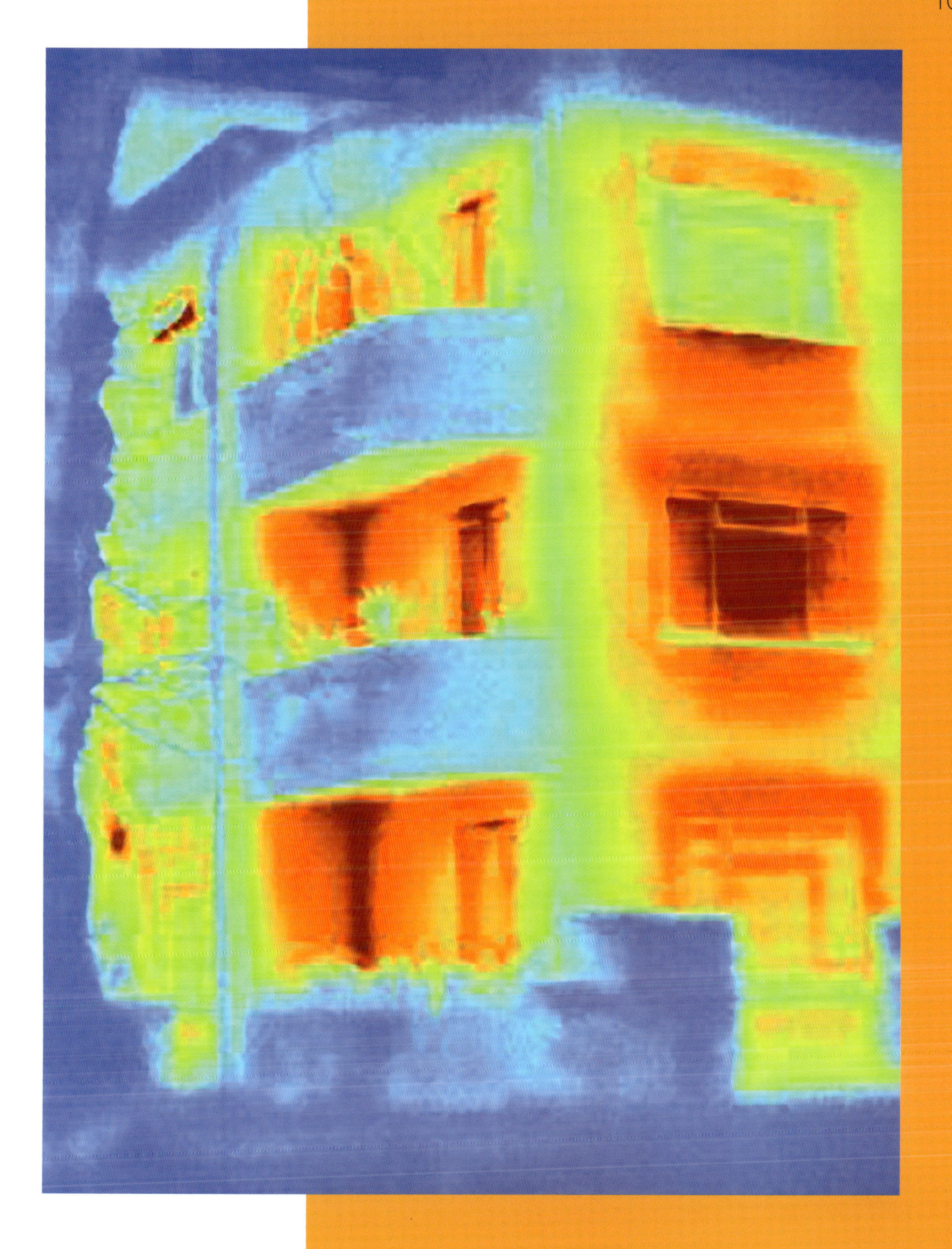

In the current age of growing ecological consciousness thermal imaging may also take on additional significance as it can help us to understand the sensory experiences of some of the creatures which inhabit the planet together with ourselves. Many living creatures possess abilities to sense heat and energy which are far superior to our own, and these abilities hugely influence their behaviour. By seeing the world through their thermal eyes we may actually come to better understand the alien environment in which our non-human companions live their everyday lives. This in turn can lead to a better understanding of our world, and even ourselves. As the old saying goes "seeing is believing", thus leveraging our sensory systems through the use of advanced technology can provide a direct route to understanding.

below: A lone figure walks down a flight of steps

opposite: Looking down over these garages and the park behind, the heat of the man below and the two lights is clearly visible

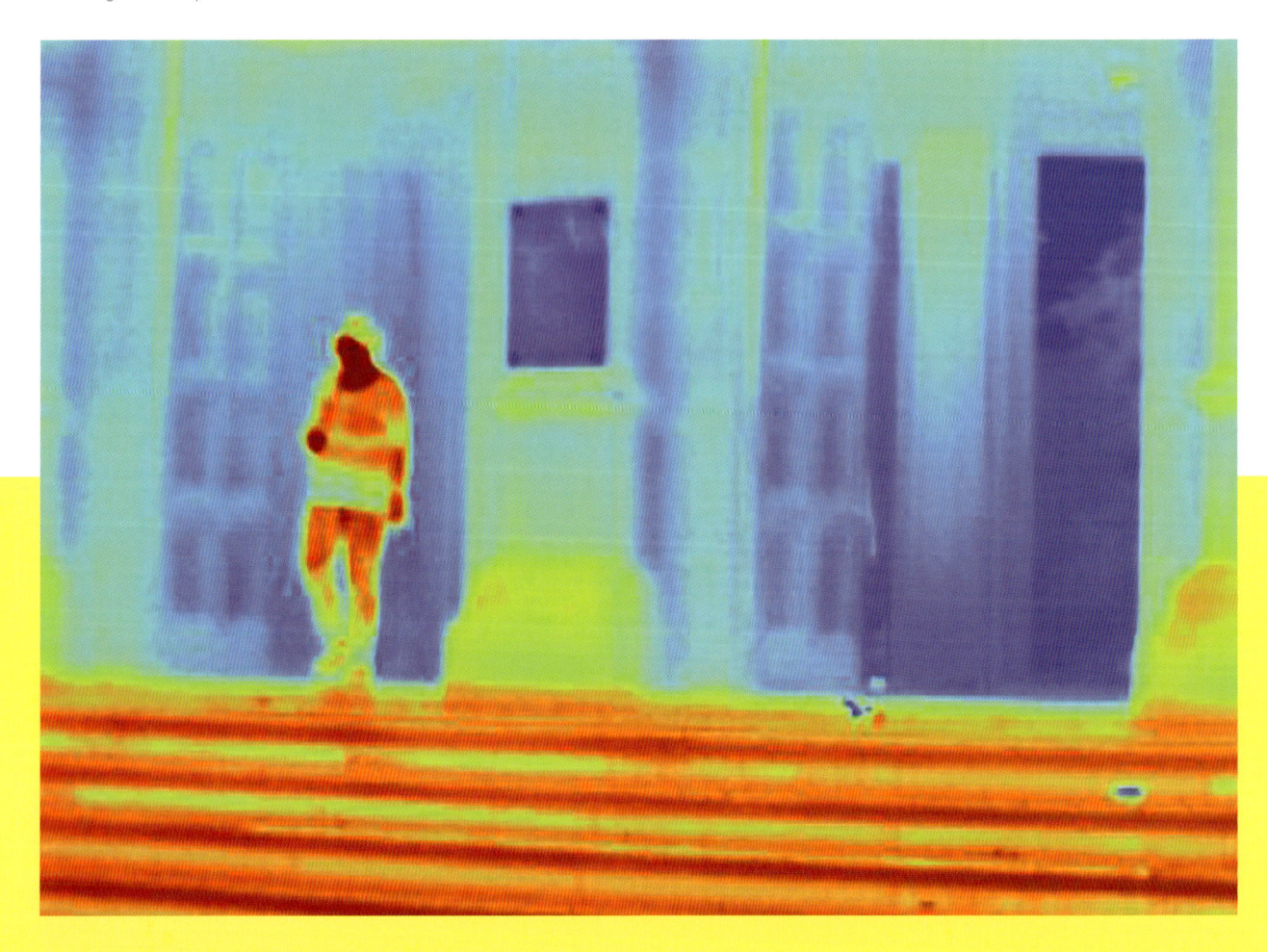

103

104

GUIDE TO THERMAL IMAGES

The thermal images found on the pages of this book are all 320x240 pixel JPEG images shot using a 60 Hz thermal imaging camera, which was similar in appearance to a camcorder. As these cameras measure a property, temperature, which is not part of the visible light spectrum, pseudo-colour is used to indicate the variations in temperature. The pseudo-colour scheme uses bright red-orange for the hottest temperature found in the individual image and dark blue for the coolest. As the pseudo-colour scheme is normalised for each image individually, the same colour can indicate different temperatures when appearing in different images. Therefore, for the current collection of images, the colour provides a measure of relative temperature rather than of absolute temperature.

Thermal imaging, or, more precisely, infrared thermography, works by measuring the infrared radiation of the electromagnetic spectrum from approximately 900 to 14,000 nanometres of wavelength. Infrared radiation is one region of the electromagnetic spectrum; other regions include gamma rays, X-rays, ultraviolet light, visible light and radio waves. Infrared radiation is emitted by all objects and the amount of emitted radiation increases with increases in the temperature of the object. The temperatures which can be measured by means of a modern thermal imaging camera are normally from approximately -50°C to 2,000°C.

Infrared radiation is measured using a thermal camera in much the same way that visible light is measured using a digital camera. However, while digital cameras use a Charge Coupled Device (CCD) or a Complementary Metal Oxide Semiconductor (CMOS) sensor, thermal imaging is based on the use of Focal Plane Array (FPA) sensors which respond to the longer wavelengths of the infrared region of the electromagnetic spectrum. Given the complexity of the FPA sensors, the maximum resolution which can currently be achieved is lower than that of CCD or CMOS sensors.

opposite:
A woman and her horse. Thermal imaging reveals that the horse's eye, mouth and nostril are the warmest parts here. By contrast, its mane, like the woman's hair, is much cooler

page 106:
A crowd of people

page 107: This close-up thermal portraits reveals which parts of the head and face are warmest

Most thermal imaging cameras have the relatively low resolutions of 160x120 pixels or 320x240 pixels, with the most expensive current models reaching 640x512 pixels.

While the amount of thermal radiation depends greatly on the surface temperature of the object which is being measured, the surface temperature is not the only factor involved. Other factors which affect the measurement include the emissivity of the object which is being shot, the amount of radiation arriving from the surrounding environment and the atmospheric absorption between the radiating object and the thermal imaging camera. Emissivity and atmospheric absorption thus affect the measured temperatures, and, if not carefully compensated at the time of each measurement, can lower the accuracy of the temperature values.

The biggest factor affecting the accuracy of thermal images is the emissivity, meaning the ability of the object's material to emit thermal radiation. Every material has an emissivity value which is in the range from 0.0 (no ability to emit thermal energy) to 1.0 (complete emission of all thermal energy). In addition, the emissivity value is not a fixed value for most materials, but is actually a continuous function of the temperature. Given the complex physics, the maximum theoretical measurement accuracy of a thermal camera is achieved only when the emissivity value of the object which is being studied is known, or when the camera can be calibrated on-site against a known reference source of thermal radiation. For the images in this book, the camera was set to run using a stored internal emissivity table, rather than calibrated for each shot, so as to achieve the maximum possible accuracy.

The thermal images presented here were selected from various photo shoots from over a period of several years, which have been grouped to help the reader best explore the strange new world of thermal vision. The photos in each group share a common characteristic, making it easier to compare content and hopefully leading to a deeper exploration of the perceptual and philosophical consequences of thermal perception.

The first twelve groups of images have been organised into what might be loosely called principles of

107

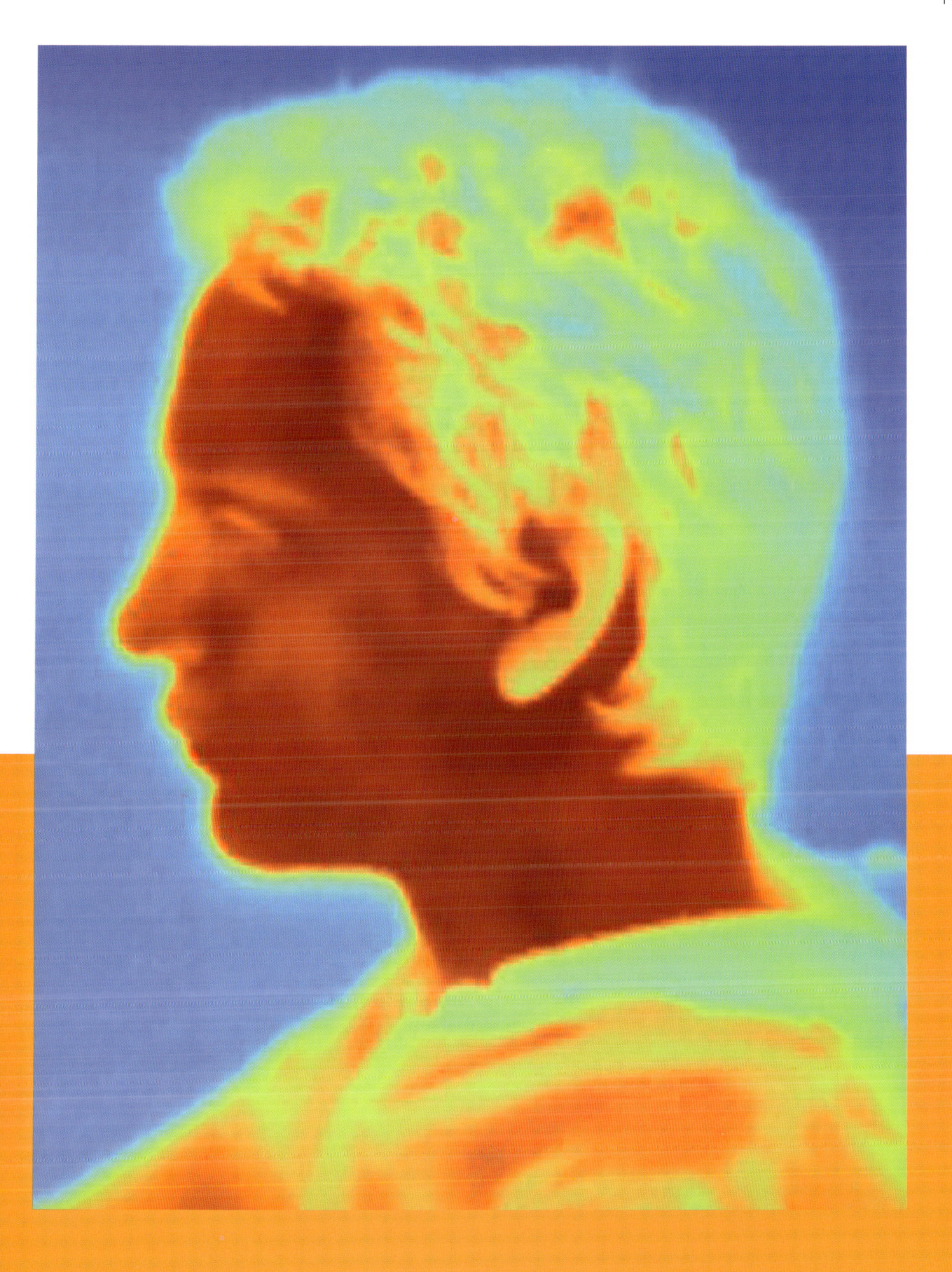

thermal perception. Any large set of thermal images will quickly reveal a few perceptual patterns which the careful observer cannot fail to notice, and these groupings attempt to illustrate some of the more obvious patterns. The second set of thermal images consists of sixteen groupings which are not based on a general principle or observation, but rather on an object or place. They feature locations or circumstances which have produced beautiful and enjoyable images whose merits often transcend the purely documentary. For example, a well-known building may be revealed in a new light thanks to thermal eyes, or something new and unexpected observed in the ordinarily mundane.

The thermal images presented in this book open a new and exciting window into the world of the "things in themselves", revealing many phenomena which we do not normally see during our everyday lives. Beyond the documentary evidence, many of the images also display expressive qualities which can perhaps move the reader, inviting him or her to reconsider some aspects of the relation between people and the natural world. It is a heartfelt hope that the images assembled in this book have enthused and excited, and that the reader will never again think of the world in the same manner as before.

below: A family walk down a busy street together

opposite: A group of people gather for discussion in this workshop

page 110-11: Two children show their fingertips, cooled from holding their iced drinks

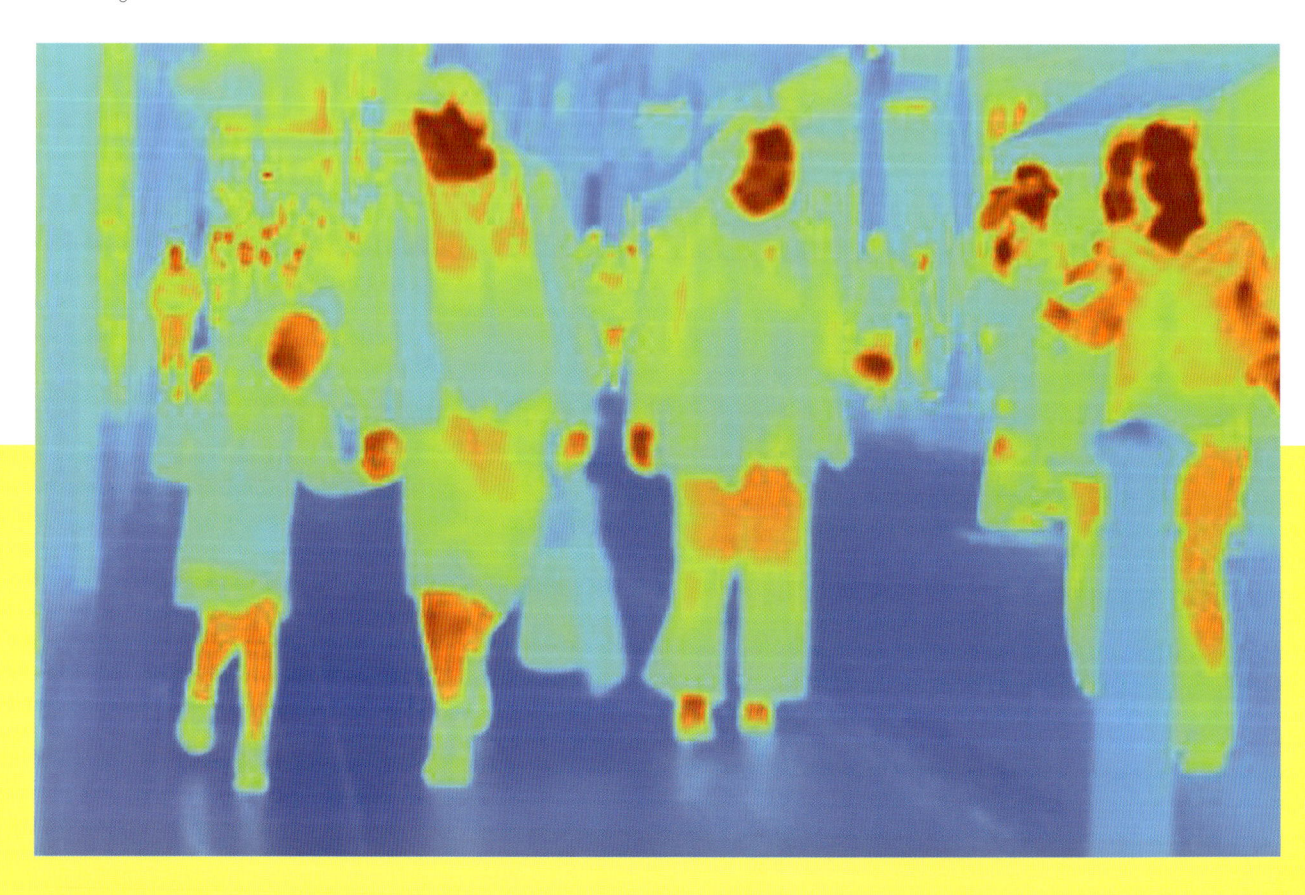

109

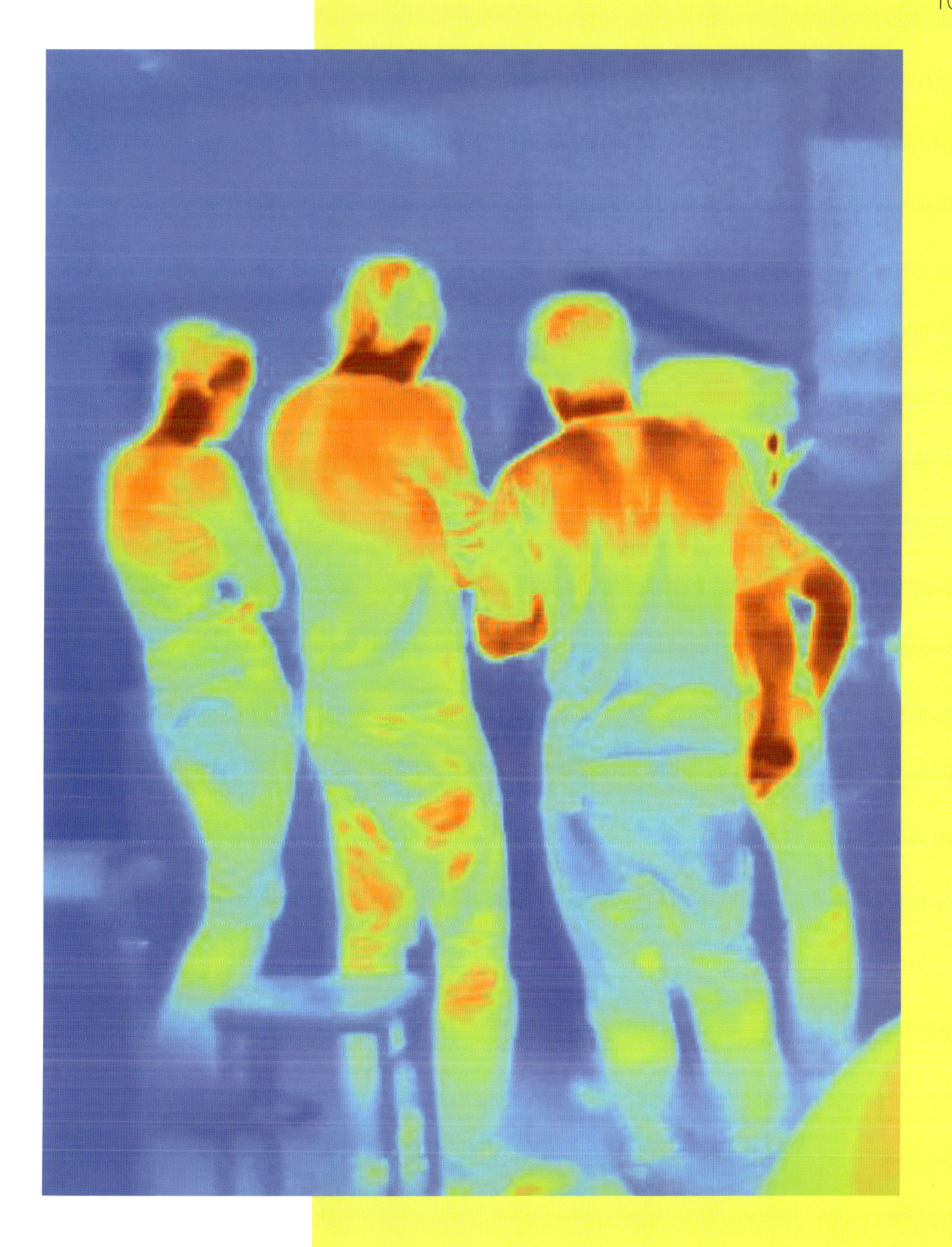

110

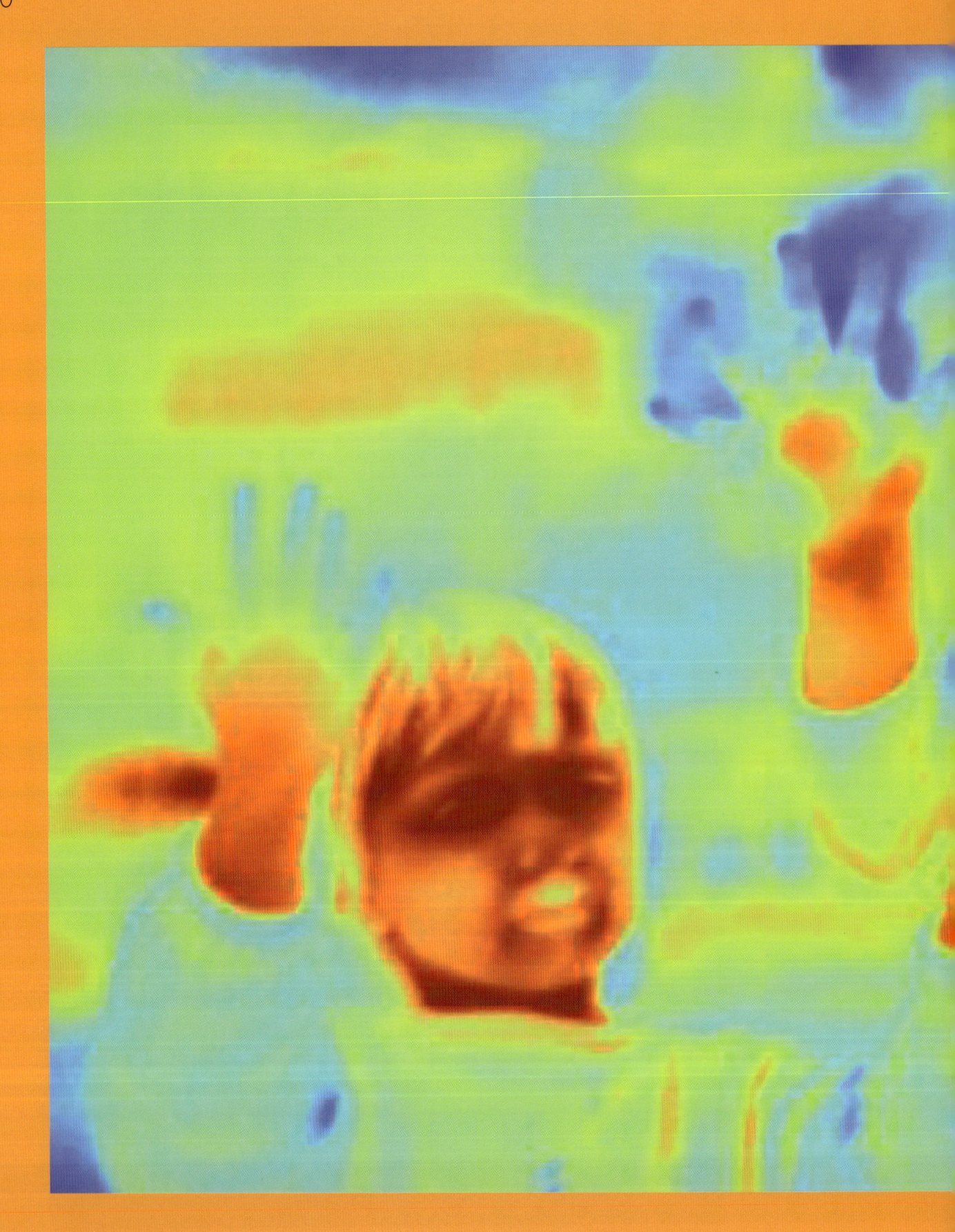

111

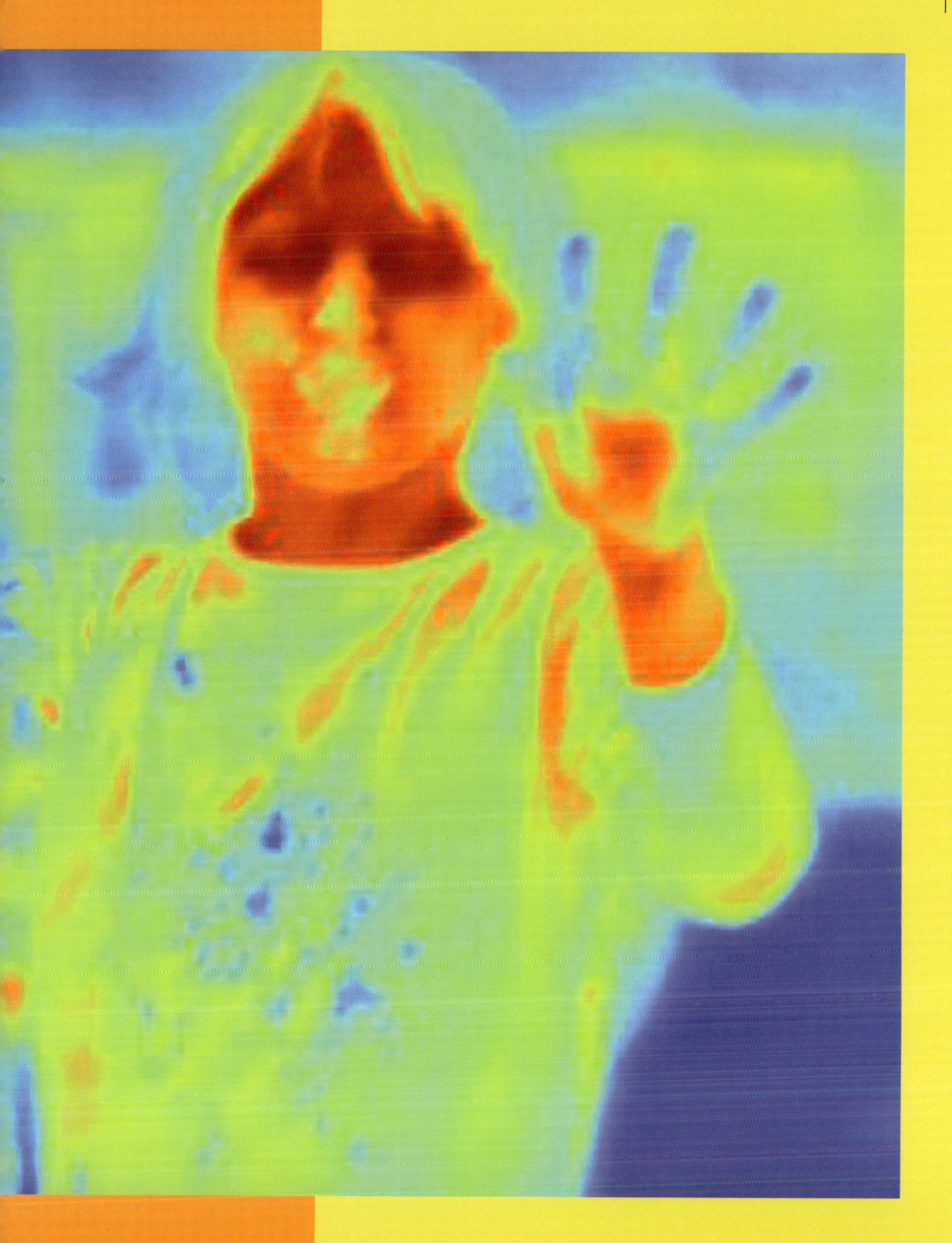

BIBLIOGRAPHY

Barnes, J. (ed.) (1984) *The Complete Works of Aristotle, Vol. 2*. Princeton University Press, Princeton, New Jersey, USA

Berkeley, G., Woolhouse R. (ed.) (2004) *Principles of Human Knowledge and Three Dialogues*. Penguin Books, London, UK

Blum, R. (1964) *Utopiates: the Use & Users of LSD 25*. Atherton Press, New York, USA

Descartes, R., Cottingham, J. (ed.) (1996) *Meditations on First Philosophy*. Cambridge University Press, Cambridge, UK

Descartes, R., Maclean, I. (2006) *A Discourse on the Method of Correctly Conducting One's Reason and Seeking Truth in the Sciences*. Oxford University Press, Oxford, UK

Diogenes, L., Yonge, C. D. (translator) (1853) *The Lives and Opinions of Eminent Philosophers*. Bohn, York Street, Covent Garden, London, UK

Gallagher, S. (2006) *How the Body Shapes the Mind*. Clarendon Press, Oxford, UK

Gibson, J. J. (1983) *The Senses Considered as Perceptual Systems*. Greenwood Press, Santa Barbara, California, USA

Griffith, R. T. H. (translator) (2008) *The Rig Veda: Complete, Forgotten Books*. Charleston, South Carolina, USA

Hartmann, G. W., Poffenberger, A. T. (1939) *Gestalt Psychology: a Survey of Facts and Principles*. The Ronald Press Company, New York, USA

Hayek, F. A. (1976) *The Sensory Order: an Inquiry into the Foundations of Theoretical Psychology*. The University of Chicago Press, Chicago, Illinois, USA

Hobbes, T., Gaskin, J. C. A. (ed.) (2008) *Human Nature and De Corpore Politico*. Oxford Paperbacks, Oxford, UK

Hobbes, T., Gaskin, J. C. A. (ed.) (2008) *Leviathan*. Oxford Paperbacks, Oxford, UK

Hume, D., Steinberg, E. (ed.) (1993) *An Enquiry Concerning Human Understanding*. Hackett Publishing Company, Indianapolis, Indiana, USA

Hunt, M. M. (2007) *The Story of Psychology*. Bantam Doubleday Dell Publishing Group, New York, USA

Huxley, A. (2004) *The Doors of Perception and Heaven and Hell*. Vintage Books, London, UK

Johansen, T. K. (2007) *Aristotle on the Sense-Organs*. Cambridge Classical Studies, Cambridge University Press, Cambridge, UK

Kant, E., Smith, N. K. (translator) (2003) *Critique of Pure Reason*. Palgrave Macmillan, Basingstoke, Hampshire, UK

Leary, T., Metzner, R., Alpert, R. (2008) *The Psychedelic Experience: a Manual Based on the "Tibetan Book of the Dead"*. Penguin Classics, London, UK

Locke, J., Woolhouse, R. (ed.) (1997) *An Essay Concerning Human Understanding*. Penguin Books, London, UK

Long, A. A. (2006) *The Cambridge Companion to Early Greek Philosophy*. Cambridge University Press, New York

Long, A. A., Sedley, D. N. (1987) *The Hellenistic Philosophers, Vol. 1*. Cambridge University Press, Cambridge, UK

Masters, R., Houston, J. (2000) *The Varieties of Psychedelic Experience, the Classic Guide to the Effects of LSD on the Human Psyche*. Park Street Press, Rochester, New York, USA

Merleau-Ponty, M., Smith, C. (translator) (2007) *Phenomenology of Perception*. Routledge, London, UK

Mooney, T., Moran, D. (2006) *The Phenomenology Reader*. Routledge, Abingdon, Oxfordshire, UK

Nagel, T. (1989) *The View from Nowhere*. Oxford University Press, Oxford, UK

Plato, Cooper, J. M. (ed.) (1997) *Plato: Complete Works*. Hackett Publishing Company, Indianapolis, Indiana, USA

Reid, T. (2002) *Essays on the Intellectual Powers of Man – a Critical Edition*. Edinburgh University Press, Edinburgh, UK

Stafford, P. (1992) *Psychedelics Encyclopedia*, 3rd edition. Ronin Publishing, Berkeley, California, USA

Strawson, P. F. (2006) *Analysis and Metaphysics: an Introduction to Philosophy*. Oxford University Press, Oxford, UK

Wasson, R. G., Hoffman, A., Ruck, C. P. (2009) *The Road to Eleusis: Unveiling the Secret of the Mysteries*. North Atlantic Books, Berkeley, California, USA

Zalta, E. N. (ed.) *The Stanford Encyclopedia of Philosophy*. Retrieved from: http://plato.stanford.edu/, Stanford University, Stanford, California, USA